I0054645

Ageing and the Visitor Econor

This significant volume is the first to use primary research evidence to examine tourism, ageing and the implications of an ageing population for the visitor economy. Adopting a multidisciplinary approach, this book offers insights into the new opportunities, threats and challenges that the growing ageing-tourism market poses.

The ageing population has created a demographic time bomb with a population structure that is skewed towards a growing proportion of older people. When this is combined with the impact of health conditions, such as dementia, the future shape of visitor demand and tourism behaviour is likely to change and face many new challenges, albeit at different rates in time and space. Chapters include cutting-edge insights into future issues, while interviews are used to illustrate and explain issues affecting ageing and tourism, creating a much-needed synthesis of the ageing–tourism nexus to demonstrate intellectual leadership around this theme.

This book will be of great interest to all upper-level students, academics and researchers in the fields of tourism, hospitality, leisure studies, and health and social care.

Stephen J. Page is Associate Dean (Research) and Professor of Business and Management at Hertfordshire Business School, University of Hertfordshire, UK. He holds an Honorary Doctorate from the University of West London and is a Visiting Professor at the University of Plymouth. He has worked in a professorial capacity for the University of Stirling, Scotland, Massey University, New Zealand, London Metropolitan University and Bournemouth University, UK, over the last 23 years. He has also worked as a tourism consultant with different organisations, including the United Nations World Tourism Organization, OECD, VisitScotland, Scottish Enterprise, Highlands and Islands Enterprise, Harrah's Casinos and Sky Tower, Auckland, New Zealand, among many other clients. He has also worked collaboratively with numerous other organisations such as VisitEngland, Historic Royal Palaces and the Alzheimer's Society. He is the author and editor of 47 books on tourism, leisure and events, and editor of the leading tourism journal *Tourism Management*. He has also been a Member of the Chartered Association of Business Schools' Scientific Committee for the Academic Journal Guide and is the co-author of the fourth edition of *Event Studies* (with Don Getz), published by Routledge in 2020.

Joanne Connell is Associate Professor of Sustainability and Tourism at the University of Exeter Business School, where she is the Programme Manager for the Master's degree in International Tourism Management. She is the author and editor of 10 books on tourism, leisure and events and is an associate editor of the leading tourism journal *Tourism Management*. She is also an associate editor of the open-access journal *Heliyon*. She has worked widely with tourism organisations and has led and worked as part of small project teams with a number of external bodies in a knowledge-transfer capacity; examples include VisitScotland, New Zealand's Ministry of Economic Development and Stirling Council. Having advised and worked extensively with Loch Lomond and the Trossachs National Park Authority (Scotland's first National Park Authority), she contributed to the development of the National Park's sustainable planning processes. Other projects include event impact evaluations, tourism business surveys, visitor surveys and organisational secondary data analysis to improve business performance. Most recently, her research has focused on helping to make the visitor economy more dementia-friendly, and she was a key collaborator with the Historic Royal Palaces and the Alzheimer's Society on *Rethinking Heritage: A Guide to Help Make Your Site More Dementia-Friendly* (2017). She also worked collaboratively with VisitEngland, VisitScotland, the Alzheimer's Society and the National Trust to produce *Dementia-Friendly Tourism: A Practical Guide for Businesses* (2019).

Routledge Advances in Tourism

Edited by Stephen Page
Hertfordshire Business School, University of Hertfordshire, UK

For more information about this series, please visit: www.routledge.com/advances-in-tourism/book-series/SE0538

Ageing and the Visitor Economy

Global Challenges and Opportunities

Stephen J. Page and Joanne Connell

Routledge
Taylor & Francis Group

LONDON AND NEW YORK

First published 2022
by Routledge
4 Park Square, Milton Park, Abingdon, Oxon OX14 4RN

and by Routledge
605 Third Avenue, New York, NY 10158

Routledge is an imprint of the Taylor & Francis Group, an informa business

© 2022 Stephen J. Page and Joanne Connell

The right of Stephen J. Page and Joanne Connell to be identified as authors of this work has been asserted by them in accordance with sections 77 and 78 of the Copyright, Designs and Patents Act 1988.

All rights reserved. No part of this book may be reprinted or reproduced or utilised in any form or by any electronic, mechanical, or other means, now known or hereafter invented, including photocopying and recording, or in any information storage or retrieval system, without permission in writing from the publishers.

Trademark notice: Product or corporate names may be trademarks or registered trademarks, and are used only for identification and explanation without intent to infringe.

British Library Cataloguing-in-Publication Data
A catalogue record for this book is available from the British Library

Library of Congress Cataloging-in-Publication Data
Names: Page, Stephen, 1963- author. | Connell, Joanne, author.
Title: Ageing and the visitor economy : global challenges and
   opportunities / Stephen J Page and Joanne Connell.
Description: Abingdon, Oxon ; New York, NY : Routledge, 2021. | Series:
   Routledge advances in tourism | Includes bibliographical references and
   index.
Subjects: LCSH: Tourism—Planning. | Older people—Recreation. | Older
   people—Travel.
Classification: LCC G155.A1 P25786 2021 (print) | LCC G155.A1
   (ebook) | DDC 338.4/791—dc23
LC record available at https://lccn.loc.gov/2021012782
LC ebook record available at https://lccn.loc.gov/2021012783

ISBN: 978-0-367-47854-4 (hbk)
ISBN: 978-1-032-07290-6 (pbk)
ISBN: 978-1-003-03935-8 (ebk)

DOI: 10.4324/9781003039358

Typeset in Times New Roman
by Apex CoVantage, LLC

Stephen wishes to dedicate the book to his parents, and especially his father, who has always extolled the virtues of getting an education as a fundamental starting point in forging one's career path.

Contents

Plates

Figures

Tables

Boxes

Preface

Ageing is a gradual but ongoing process that begins as soon as we are born and continues as we progress through different stages of human development. In the latter stages of the human life cycle, many of our bodily functions begin to decline (Schäfer 2011). Hall (1922) conceptualises this as *senescence* – deterioration that comes with old age. Even so, people do not start to age at a predefined point in time. Despite this, many societies have thought of the traditional retirement age of 60–65 as the beginning of old age or the era in which the process of becoming elderly begins, although this is of course a widely contested subject in the social sciences and science. Ageing as a subject of study is a comparatively recent phenomenon of major interest within the social sciences, although the study of ageing can be traced to the early nineteenth century and Quetelet's (1836) *Sur l'homme et la développement de ses facultés, ou essai de physique sociale*. But the development of gerontology, which broadly defined is the scientific study of ageing and the problems which the ageing population face, can largely be attributed to its evolution in the early-to-mid twentieth century, as illustrated by some landmark studies (e.g. Hall 1922; Kaplan 1946; Shock 1952; and the publication of the first issue of the *Journal of Gerontology* in 1946). One of the most obvious reasons for the lack of attention devoted to ageing as a field of academic of endeavour was the relatively short lifespan in industrialising nations in the late eighteenth and nineteenth centuries. Robert and Labat-Robert (2017) illustrated this with reference to Paris, where residents' life expectancy increased by 256% between 1750 and 2016. This could be attributed to a number of changes that took place over this period: first the prevalence of disease and poor nutrition were addressed, followed by the rise of vaccinations, and, especially after 1950, the reduction of cardiovascular disease; all of these measures affected factors that had hitherto suppressed life expectancy. Similar patterns can be seen in many other developed countries over the same period, and more recently in many emerging industrialising countries.

In many respects ageing was an invisible social policy issue in the industrialising countries of the nineteenth century, prior to mid-Victorian state interventions in many countries to curb the excesses of capitalism, where minimal state support and charities dealt with the poverty associated with ageing as poor relief. Such relief often had a moral undertone, making the distinction between 'deserving' and 'undeserving' poor, a theme that continued through to the establishment of universal social support for the aged in the twentieth century. Several strands of state intervention to improve the condition of the poor in Victorian cities (Briggs 1963) emerged, which incrementally contributed to extending the lifespan and life chances of the population through improved sanitation, clean water supplies, better housing, the provision of open space and increased leisure time in the later nineteenth century. As Boyer and Schmidle (2009: 250) poignantly illustrated,

> Victorian Britain had a large elderly population. In 1861, there were 932,000 persons aged 65 and over in England and Wales, representing 4.6 per cent of the population. By 1891 there were nearly 1.4 million persons aged 65 and over, or 4.7 per cent of the population, and 800,000 aged 70 and over. A person aged 35–9 in 1861 or 1871 had about a 50 per cent chance of surviving to the age of 65–9. The life expectancy of a 65-year-old male was 10.7 years in 1861 and 10.6 years in 1901; that of a 65-year-old female was 11.6 years in 1861 and 11.8 years in 1901.

The prevalence of an 'ageing population' as a societal issue requiring specific forms of state intervention periodically surfaced in policy debates around the condition of the poor, most notably in the UK in several social surveys (Booth 1889; Rowntree 1901) and a Royal Commission (Royal Commission on the Aged Poor 1898) and with the provision of pensions for the over-70s in the Old Age Pension Act (1908). Critiques of the early reports into ageing noted a clear link between old age and poverty (Yule 1899; also see Hepple 2001), since 'incapacity for work resulting from old age' (Royal Commission on the Aged Poor 1898: vi) drew attention to the ideology of ageing and state dependency. These attitudes changed as the social construction of ageing developed as a distinctive theme in the Victorian and Edwardian age, and as old age began to be thought of as a stage of life in its own right, as a result of the actual numbers of people who were living longer (Chase 2009; Cole 1992). Townsend's (1981: 9) poignant assessment that 'society creates the framework of institutions and rules within which the problems of the elderly emerge, and indeed, are manufactured' is a reminder of how ageing has been recognised as a societal problem in need of state intervention over time, and state responses through the provision of pensions and institutionalisation of retirement emerged in the twentieth century. Nonetheless, Townsend also recognised that ageing in the post-war period continued to trigger an institutionalised response by the state and also to be viewed as a major societal problem, as it continued to be associated with poverty in old age for some. At the same time, for others, it was possible to age while enjoying the comfortable and leisure-laden lifestyle driven since the 1960s by the consumer society seen in developed nations, and which has been fuelled by private and state pensions and wealth among certain groups.

Ageing continued to be a topic of interest globally in many countries, as the United States demonstrated with its First National Commission on Ageing (United States Federal Security Agency 1951). The UK equivalent was the Royal

Commission on Population (1949), a report which forecast considerable growth in the over-65 age group in the post-war period, especially with the interventions of the newly formed National Health Service, which continued to extend life expectancy through free healthcare provision. Other notable reports such as Rowntree (1947), Thompson (1949) and the Phillips report (HMSO 1954) identified a series of policy-related issues on ageing that were common to many developed countries. Many decades later, these ongoing concerns were revisited in reports published by the Royal Commission on Long Term Care (1999) and the House of Lords (2012).

The polar opposite of this recognition of old age as a societal problem was apparent in the link between tourism and ageing that can be dated to the eighteenth century, with the evolution of spa tourism linked to health and well-being among the 'leisured classes' (i.e. those who were affluent enough to travel). Diarists and commentators identified the benefits of 'taking the waters'; see for example Currie's (1793) article, which helped to form the basis of hydropathy as a curative leisure pursuit (Durie 2006) among the aged infirm, as reflected in the age profile of spa visitors (Rotherham 2014). As Blaikie (1999: 136) argued, the 'eighteenth century certainly seems to possess a "therapeutic nihilism" with regard to old age as a whole. Yet it also demonstrates a tendency to carve up senescence into various ailments', and the mythical health benefits of visiting spas and taking the waters were widely advocated, albeit on the basis of spurious scientific evidence and recommendations. In no way did this constitute a scientific analysis of old age or healthy ageing, but for a certain class these pursuits were very popular, as is evident from the excellent collection of source materials on the subject in the British Library (British Library n.d.). This collection includes examples of how Bath and the cures proposed by 'quack doctors' were documented and parodied in cartoons and novels which mocked the pretensions of polite society and its leisure habits. These texts also illustrate a significant and enduring theme in tourism research: its focus on affluent and ageing visitors (e.g. Macpherson 1869). At the global scale, Walton (2014) draws parallels between different countries with regard to this form of tourism, some of which was associated with opportunistic marketing and the acquisition of social and cultural capital and status from visiting such locations. The principal theme in the early association between ageing visitors and visiting spas and the coast was the affluence and leisure time that were needed to engage in these activities. Move forward 300 years and the difference is that there has been a democratisation of opportunities for travel and leisure in the post-war period. The first generation to benefit fully were the 'baby boomers' (born between 1944 and 1964). They enjoyed greater domestic leisure and tourism mobility due to increased car ownership (Patmore 1983) as well as the unprecedented overseas travel opportunities afforded by the rise of the package holiday (Bray and Raitz 2001; Inglis 2000; Löfgren 2002; Walton 1983, 2000) and cheap air travel. These experiences eventually prompted a more critical review of the relationship between ageing and tourism, as the baby-boomer generation became seasoned travellers as well as the main beneficiaries of post-war wealth and affluence. As noted earlier, there are exceptions to this rule, with poverty and deprivation still prevalent in the post-war era, despite periods of full employment in many developed nations, as capitalism continued to perpetuate inequality in society (Townsend 1979, 1981).

Many of the experienced travellers of the baby-boomer generation are now well into their seventies, and they can expect to live (and possibly travel) for several more decades yet. This may well bring considerable challenges as well as opportunities for the wider visitor economy in terms of both diversity and volume of travellers. Against this background, it is timely to consider the ageing–visitor economy nexus if we are to gain a richer understanding of the implications of demographic change for destinations, tourism businesses and the travelling public itself.

Acknowledgements

The authors would like to thank a range of people who have made this book possible. Stephen wishes to thank Professor Damian Ward, Dean of Hertfordshire Business School, for his support in granting Stephen a period of sabbatical towards helping to complete this project in 2021 as well as financial support towards transcription and survey costs. In addition, Julie Franklin helped transcribe many of Stephen's scribbles and notes from a number of different sources, including the British Library, into a coherent text. Jordan Parker used her excellent skills in transforming some rough diagrams into publishable content. Dr Tetiana Hill provided her usual cheery and helpful input when undertaking the interviews during a national lockdown. Finally, at Routledge, Stephen thanks Emma for her willingness to commission the title and Lydia for handling the manuscript.

Joanne wants to express her gratitude to the many organisations and individuals who have assisted this project over the years, including the countless people we interviewed in 2021. They are too numerous to list here, but they all helped us keep up the momentum by kindly sharing their views and insights.

Finally, we would like to thank the ESRC for supporting our ongoing work on people with cognitive decline through the 'Extending active life for older people with cognitive impairment through innovation in the visitor economy of the natural environment' grant (ENLIVEN; ES/V016172/1), which will help us to progress many of the issues we discuss in the book.

Copyrighted material

The following material is reproduced with kind permission from the copyright owners:

- Routledge for Figures 1.2, 2.1, 2.5, 3.4, 4.4 and 7.2 and Tables 2.7, 2.8, 4.1 and 4.2;
- Elsevier for Figures 2.2, 3.1, 3.2, 5.1, 5.2 and 7.4; and
- Pearson Education for Figures 4.1, 4.3 and 6.1.

1   Introduction

Introduction

In February 2020, world media attention was drawn to the quarantined cruise ship *Diamond Princess*, which was docked in Yokohama, Japan, as its owners sought to bring a Covid-19 outbreak under control among its 2465 passengers and 1068 crew (Mizumoto *et al.* 2020). The passenger profile was typically over 55 years of age with a median age of 69 and generally in the range of 60–74 (Yamagishi *et al.* 2020). This unfortunate event, one of the key events in the early stages of the pandemic, went some way to emphasise the importance of older travellers as a core market in international travel and tourism. As the pandemic advanced, our ageing population took centre stage, not only for the predisposition of the virus to cause higher rates of mortality among the over-75 age group, but for the loneliness and isolation that adversely affected older people during lockdowns. This age group accounted for 75% of all deaths in the UK and 60% of all deaths in the USA up to January 2021. Legal restrictions and closures, social distancing (Plate 1.1) and safety provisions when travelling (Plate 1.2) led to reduced contact with family and friends, and steep declines in a wide range of tourism and leisure-related activity.

Despite the immediacy of the impact of the pandemic on the leisure activities of older people, ageing is not a new theme within the field of leisure and tourism research. One of the most obvious examples is the interest in retirement migration to areas associated with leisure and tourism environments in older age (e.g. the coast) and its subsequent development in relation to research on mobilities. Early studies can be traced to the 1950s and 1960s (e.g. Hitt 1954; Mercer 1970; Gilbert 1965; Harrison *et al.* 1971; Wolfe 1966; Karn 1974, 1977; Law and Warnes 1973, 1976; Warnes and Law 1984; Williams 1998), and similar research has continued into this century (e.g. King *et al.* 2000; Warnes 2001; Williams and Hall 2000; Williams *et al.* 2000). Formative studies within the area of ageing, place and space (e.g. Golant 1972, 1984; Rowles 1978, 1986; Warnes 1981, 1982, 1990, 2009; Skinner *et al.* 2014, 2017) and population geography highlighted the ageing migration–leisure/tourism nexus, as coastal and inland resorts developed a skewed localised population structure dominated by retirees (e.g. Mellor 1962;

DOI: 10.4324/9781003039358-1

Plate 1.1 The impact of coronavirus on travel – mandatory social distancing measures

Plate 1.2 Measures to reduce the spread of coronavirus when travelling – mandatory mask use and intercity rail travel

Rogers 1974; Law and Warnes 1973; Elvidge 1973; Duffield 1984). Despite some attempts to depict tourism as a form of migration (e.g. Wolfe 1966), other studies have clarified the distinction between tourism as a temporary form of migration and permanent migration (e.g. Bell and Ward 2000). Williams and Hall (2000: 7) demonstrate the link between tourism and migration where

> Many forms of migration generate tourism flows, in particular through the geographical extension of friendship and kinship networks. Migrants may become poles of tourist flows, while they themselves become tourists in returning to visit friends and relations in their areas of origin. These ebbs and flows of tourism are structured by the life course of the migrants, with each temporary or permanent round of migration creating a new spatial arrangement of friendship and kinship networks, which potentially represent visiting friends' and relations' tourism flows.

To this we need to add the daily lives of an ageing population and their leisure activities, notably the latter stage of one's life course.

Yet the history of tourism scholarship shows that scant attention has been paid to the interconnections between tourism, leisure and the broader area of demography, from which many early studies of retirement migration emanated. It is a curious and a persistent omission in tourism research that the leisure/tourism–demography nexus is overlooked. Age is a major consideration in mobility throughout the human life course, as barriers and constraints develop with age. As a result, we need to look more closely at the emergent literature on mobilities (e.g. Sheller and Urry 2006; Hall 2015; Adey *et al.* 2014; Janta *et al.* 2015) because it offers more theoretically informed perspectives on how tourism and leisure mobility are part of a broader concern with human movement and the mobile daily lives of people. Indeed,

> Demography is one of the external factors that shape tourism demand and development. The structure of societies is continuously changing, and for both public and private organisations working in the tourism field it is relevant to study these changes in order to anticipate and react to them in the most competitive way. Demographic changes are likely to impact on the patterns of travel demand, including frequency, length of stay, products, and consequently the communication strategies of National Tourism Organisations (NTOs) and private companies alike.
>
> (UNWTO 2010: 1)

Ageing is a key element of demography, with its focus on the components that influence its growth and decline (i.e. births, deaths, marriages/civil partnerships) and the changing structure of population through time, including forecasts of future changes in its composition. Given that demography plays a key role in the planning and policy development of countries, regions and localities, it is surprising that so few studies have explored the demography–leisure/tourism nexus as a strand of leisure and tourism research since its early development as a social science subject. The only notable exception to this is Yeoman *et al.* (2010). This benign neglect of the demography–leisure/tourism nexus, aside from studies of migration and temporary mobilities as tourism (i.e. outcomes of the movement of people from their home to their destination and back), means that the fundamental tenets of demography that condition and influence the mobility potential of the population have been overlooked. The demographic structure of a specific area, region or country is an implicit assumption in most studies of tourism even though it shapes the demand for tourism and leisure in its most simplistic formulation. Instead, different disciplines, most notably marketing, analyse and seek to explain the demographic outcome of tourism and leisure activity and behaviour through the lens of consumer behaviour or from a social psychological perspective of motivation rather than from a demographic perspective. Such studies observe the changing nature of motivations and consumer behaviour, recognising that the population structures of many tourism-generating regions (i.e. the source of demand) are – or will be – influenced by ageing. But this does not explain the demographic blindness of many tourism (and to a lesser degree leisure) research-ers, presenting a puzzle when looking at any of the seminal works on tourism or the numerous texts on tourism: if tourism is about people, then the subject matter must be primarily embedded in demography, yet this truism is routinely overlooked. In short, demography is the basis of the 'people' element in all forms of tourism and leisure activity, from the demand for goods, services and experi-ences to their supply by the workforce that provides the labour for what is widely acknowledged as a people-intensive industry.

Consequently, a fundamental understanding of demography would seem to be apposite if we are to appreciate fully the issues that will shape the future visitor economy. As one component of the demography of any country, region or local-ity, ageing is arguably the greatest challenge that will affect the visitor economy over the next decade and beyond, given the global forecasts that tourism will reach 1.8 billion arrivals by 2030. In simple terms, by 2030, a greater proportion of world travellers and tourists will be older than we see today, so the profile and structure of the travelling public will have transformed, as can be inferred from studies dating back to the 1990s (e.g. United Nations Division for Social Policy and Development 1998). Equally, daily leisure demand will be dominated by the challenge of ageing. Box 1.1 identifies a broad range of reasons why leisure/tour-ism, ageing and the visitor economy must be understood at both the global and the local scale, since ageing will change the drivers of leisure demand and thus require the creation and redesign of a different set of services and experiences to accommodate a different consumer profile.

Box 1.1   Key reasons to study the ageing population–tourism nexus

Demographic and societal considerations

- Increasing life expectancy of the population in developed and devel-oping countries due to eradication of diseases (e.g. vaccinations) and improved living standards, education and healthcare (e.g. reductions in traditional causes of morbidity, such as heart attacks).

- Rise of more complex and modern health concerns among the ageing population (e.g. diabetes, dementia, sight loss, deafness and treatable diseases).
- Changes to the nuclear family structure, and isolation and loneliness as socio-psychological issues among an ageing population with more fragmented families.
- Greater dependence upon healthcare systems for maintaining longer life expectancy.
- Funding issues faced by governments and their care systems as they attempt to provide services for an ageing population.
- Increased role of the third sector in advocating the rights of an ageing population and their care.

Visitor economy considerations

- An increasingly 'invisible' sector of the visitor economy (e.g. Barclays 2015) despite the economic spending potential of the group in terms of leisure and tourism activities both individually and with other generations (e.g. children and grandchildren).
- Growing filial responsibility among adult children and the impact of this on holiday taking and leisure behaviour.
- Ageing travellers account for an increasing proportion of holiday taking and leisure consumers in developed and developing countries.
- Baby boomers (born between 1946 and 1964) are at the peak of their consumer spending now, reflected in the expansion of this market segment.
- Ageing travellers taking different types of holidays as they seek more memorable or educational experiences, soft adventure activities (e.g. trekking and guided trips) and/or volunteer work.
- The rise of older 'new age' travellers (Schiffman and Sherman 1991), reflected in a desire to seek new, more spiritual and experiential activities, such as cultural and heritage travel combined with a focus on shorter but more frequent visits as well as off-season long-haul trips. The members of this group tend to be health-conscious and their thinking is usually much younger than their chronological age.
- More technologically savvy consumers with large spending potential in some market segments due to pension benefits coupled with house price inflation and home ownership.
- A wide range of tourism experiences dating back to the late 1950s and 1960s corresponding with the boom in package holidays means that older travellers are highly experienced, having visited a wider repertoire of destinations than the previous generation.
- A resurgence in active leisure lifestyles, dating from the 1930s, which has been reformulated for an ageing population in terms of active and successful ageing to address their needs for exercise and well-being.

Source: Developed by the authors from Patterson (2018)

Subsequent chapters will develop the complexities of the themes outlined in Box 1.1. In the meantime, this chapter will explore a number of the broader contextual issues that help to explain why ageing has assumed global significance as a societal and business issue that cannot be ignored. We start with an overview of the research questions that we sought to answer, followed by the methods we employed to generate primary and secondary data to address these questions. Next, we explore ageing as a grand societal challenge, the global dimensions of ageing and how ageing has evolved as an interdisciplinary research area. This leads on to how we should conceptualise ageing as a focus for Chapter 2. Attention then shifts to the main themes of the book, and the rationale for extending the focus beyond the narrow confines of tourism to consider the much broader visitor economy.

Research questions

The primary aim of this book is to create an innovative, thought-provoking synthesis of the burgeoning literature from social science and science on the theme of ageing, and from that synthesis to draw out the principal issues as they relate to the visitor economy. This is the first time that such an all-embracing review of the ageing–visitor economy nexus has been attempted from a broad social science/ science perspective, since the typical approach has been to focus inwardly on the subject area (e.g. tourism or leisure) and to approach the synthesis predominantly from that perspective. In this book, we adopt a very different approach in which we address at least three broad and interconnected research questions as the basis on which to structure our main thesis:

- What is ageing from a social science and science perspective?
- What is the significance of the ageing population as a global process and what are the implications for the visitor economy?
- What measures can be taken to develop a more age-friendly visitor economy and how can these be best implemented?

The book addresses these questions within one integrative framework – the ageing domain of knowledge (Figure 1.1), which validates the scope of the synthesis required to create a detailed understanding of ageing as a construct. The purpose of the synthesis is to create a simplified narrative that links together many of the critical issues with a view to exploring the interconnections and intersections between disciplines, subject areas, issues and practices as they relate to ageing and the visitor economy. The synthesis is never going to be entirely complete, but it seeks to cover the most important ground around ageing so that the uninitiated can gain a deeper understanding of the ageing–visitor economy nexus in one accessible source.

Methodology

We have employed a range of methods to look at the ways in which ageing has evolved as a societal construction, where it is often simplistically labelled as a 'problem' or a 'challenge' due to the changing dynamics of demography at both

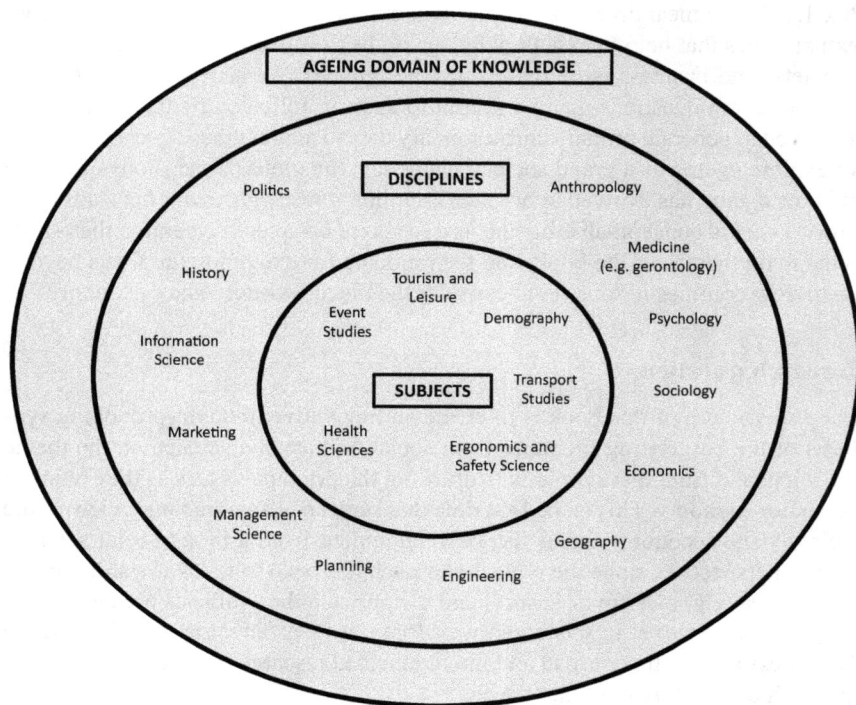

Figure 1.1 The ageing domain of knowledge

the global and the national scale. Our research philosophy adopts a pragmatic approach that recognises the findings of different methods to understand the complex relationships that exist between the ageing population, the structure of society and its institutions, and the underlying political and economic ideologies that seek to manage those relationships, including the visitor economy. We rely upon mixed methods (Teddlie *et al.* 2020; Creswell 2014), as our own empirical analysis and the studies we draw upon are largely exploratory. Integrating a wide range of research findings in such a way will aid understanding of the agendas, challenges and issues associated with the research questions we have posed. Moreover, because ageing is a complex social phenomenon it does not necessarily lend itself to a specific approach to understanding the multitude of issues and interconnections with business-related themes in the visitor economy. There are layered experiences of ageing at both a societal and an individual level, and we argue that a more pragmatic approach, which draws from interpretivist and positivist traditions, will aid a greater understanding of the relationship between ageing and the visitor economy. Creating a synthesis of the existing knowledge base alongside an analysis of primary research data to address our research questions also means that we attach a great deal of importance to emergent approaches such as critical gerontology, given recent interest in this subject:

> the connection between critique and gerontology rests upon the possibilities for creating a kind of society that better incorporates the needs and interests of older people. Critique may play many roles in this regard, beginning with opening up the possibilities of later life by paying close attention to the everyday lives of older people.
>
> (Doheny and Jones 2021: 18)

This underlines one of the principal objectives of this book – to understand how the visitor economy might better recognise the needs of an ageing demographic and meet those needs – which presupposes that social theory is a powerful explanatory tool when looking at the interconnections of ageing and the visitor economy. Unfortunately, though, existing research on the visitor economy is still highly atheoretical (Page and Connell 2020), especially when compared to analyses of ageing in disciplines such as economics, sociology and geography. This is highly problematic when looking at a cross-cutting theme such as ageing and the visitor economy, which is why we have adopted a more applied approach as opposed to a highly theorised one. As a result, this monograph is primarily an exploratory investigation.

The following methods are used to address the research questions:

- Establishing, surveying and synthesising the knowledge base that forms the parameters of the ageing domain of knowledge (Figure 1.1), which generates a diversity of perspectives that have shaped the modern-day analysis of ageing.
- Reviewing the grey literature on ageing (e.g. unpublished or published research disseminated non-commercially, such as government reports, policy statements and other position papers), which is now readily available on the internet.
- Reviewing best practice cases on the management of ageing and the provision of services and experiences to understand progress and practice on connecting ageing and the visitor economy.
- Using primary research findings comprising content analysis of websites/ information and service provision for ageing travellers to understand the current state of development around ageing and the visitor economy.
- Using primary data that provides a combination of tacit and explicit knowledge that the authors developed for an impact case study on behalf of the UK's Research Excellence Framework in 2021 around the issue of developing dementia-friendly business practices.
- Reporting primary research findings of semi-structured interviews conducted during 2021 with 20 stakeholders associated with ageing policy and practice with a view to demonstrating the wide variety of approaches to ageing as a grand societal issue, particularly in terms of the visitor economy. These interviews were designed to enhance the discussion and ground some of the academic literature in a more pragmatic context, given the richness

of the participants' qualitative comments. The conversations were entirely exploratory in nature and based on a number of questions (see Appendix) that were designed to aid understanding of the relationship between ageing and the visitor economy by extending existing research on cognitive decline and dementia as well as ageing more generally. Only a small proportion of this source material appears here, and only where it is wholly relevant; it does not form the basis of the entire monograph. The interviewees were sampled from a range of international and national organisations and businesses, and their anonymity was guaranteed and scrupulously maintained to encourage complete candour. The interviews were undertaken at a particularly challenging time for research on account of the Covid-19 pandemic and various lockdowns, so we are particularly grateful for the valuable insights the interviewees provided.

Methodological issues associated with the interviews

An interview schedule was drafted to extend some of the ideas we had developed in previous studies of dementia (Connell *et al.* 2017) to facilitate a deeper investigation of business awareness and attitudes towards ageing issues (see Appendix). The central tenet of the interviews was to garner insights from a range of stakeholders, including: those who are already involved in or are developing ageing markets; those who have a strategic/policy overview in the sector; and those who have more specialist knowledge of specific aspects of ageing. A sample of respondents was selected using the key informant technique to identify a range of key businesses, service providers and organisations within the sector, which was extended through snowballing procedures. A total of 20 interviews were undertaken by telephone or video conference in January and February 2021. These were audio-recorded and subsequently transcribed along with the emailed responses from those respondents who had chosen to supply information via a semi-structured questionnaire with an identical range of questions. Interview length ranged between 45 and 60 minutes and a large volume of information was generated, reflecting both the sensitivity and novelty of the topic. While the questionnaire schedule acted as a structure for the initial analysis, an a priori and emergent thematic approach was adopted to construct sub-themes and codes.

The interviewees were drawn from the following businesses:

- Two tour operators with an international focus, operating in the UK and Eastern Europe;
- One UK-based visitor attraction with a profile of ageing consumers;
- One UK-based charity and one European-based charity focusing on a specific disability; and
- 13 age-related charities with either a national or a global focus, operating in the UK and Eastern Europe.

The coding process comprised several in-depth readings of each interview transcript followed by grouping of key issues, as advocated by Ritchie and Lewis (2003).

The transcripts were subject to open coding on the basis of a priori and emergent themes, which have informed the organisation of this book. Axial coding to refine groupings of issues in relation to interview narratives, and to confirm relationships between categories, was then applied to highlight the primary issues. Finally, selective coding was applied to identify patterns in the data and possible connections between responses (see Corbin and Strauss 2014). Short quotes from interviews are embedded throughout the text to provide evidence on key ideas and themes and to represent a range of views, as advocated by Creswell (2013) and Merriam (2009).

Ageing as a grand societal challenge

The emergence of an ageing population structure is a global phenomenon, observed especially but not exclusively in advanced developed countries, and is one of international importance. According to the UK's Department for Business, Energy and Industrial Strategy (2017: 55),

> The UK population is ageing, as it is across the industrialised world. The prospect of longer lives will require people to plan their careers and retirement differently. Ageing populations will create new demands for technologies, products and services, including new care technologies, new housing models and innovative savings products for retirement. We have an obligation to help our older citizens lead independent, fulfilled lives, continuing to contribute to society. If we succeed, we will create an economy which works for everyone regardless of age.

The overall aim is to 'ensure that people can enjoy at least five extra healthy independent years of life by 2035, while narrowing the gap between the experience of the richest and poorest'.

This summation from a UK context illustrates a number of key themes that have much wider applicability, such as the following:

- Many developed countries expect their demographic profiles to become more aged, despite the impact of global migration. For instance, in the USA, it is suggested that the proportion of people over 65 years of age will increase from 15% of the population in 2017 to 21% by 2030 (a higher figure than those aged 16 years and under).
- Work and employment patterns are changing too, with some people already working well beyond the traditional retirement age. The ING (2020) report recently highlighted the challenges and opportunities of an ageing society when suggesting that demographic change and a shrinking workforce of 18–65-year-olds mean companies will need to retain older workers for longer through a variety of support mechanisms.
- The over-65 age group was reported as having a £3 trillion spending capacity in Europe and globally, and the over-60 age group had a US$15 trillion spending capacity. Yet the ING (2020) emphasised the importance of adopting a nuanced approach to ageing in terms of workforce and consumer spending and the need to adapt to new realities as strategic growth opportunities emerge.

- The ways in which older people live, consume and are cared for are all likely to change, with conventional thinking on institutional care (Townsend 1981) now challenged in debates about living independently rather than in a dependency relationship with the state.
- Governments, public health systems and pension providers will all need to plan for increased life expectancy as populations continue to age and become more aware of their position in society and more technologically savvy. New, yet to be developed technological solutions may prove helpful in advocacy of ageing population needs. Conversely, increased longevity will bring significant challenges for healthcare and other sectors of society (e.g. later-life diseases and social support).

Many governments now accept that their ageing populations constitute an explicit and well-publicised 'grand challenge' (e.g. Department for Business, Energy and Industrial Strategy 2017). Rapidly changing demographic profiles certainly demand a fundamental rethink of how society is organised and how it should support ever-longer active (and inactive) lifestyles, including by enhancing mobility and travel to achieve greater accessibility. However, political short-termism has meant that many of the same governments have introduced only minor improvements to provision and financial support for an ageing population while attempting to deflect public pressure for more substantive action by ordering a series of public inquiries into the issue.

In 2009, Crampton noted the start of a 'longevity revolution' that was leading some developed nations to categorise 'old age' as over 80. Crampton extends the analysis of ageing as a grand challenge that faces both the developed and the developing world. He continued:

> Today older adults may seem to have a greater presence in wealthy nations of the global North because the proportions of people 60-plus are higher … However, most people aged 60-plus and about half of those aged 80-plus live in the global South. By 2050, 69% of the oldest old will be living in that region … While the population of people 60-plus will increase from 231 million to 395 million between 2000 and 2050 in the global North … the older population in the global South will increase from 374 million to 1.6 billion during the same period.
>
> (Crampton 2009: 4)

As Crampton (2009) argued, concerns about the global population explosion in the 1950s have now been replaced by anxiety over an ageing population and age dependency. This dependency affects the types and models of service provision that the ageing population will need. It illustrates that we need to accept that ageing is a grand challenge and a pivotal issue for society and the economy. Narrow conceptualisations of ageing must be overcome, and policymakers must move beyond 'how to manage and care for old people who are unproductive and dependent … [and recognise] that ageing is part of the entire lifespan and not simply a loss suffered around age 60' (Crampton 2009: 34). Part of the major rethink that is needed relates to the way society has normalised the concept of ageing and perceived the ageing population as an invisible, passive and largely powerless element (aside from during election campaigns). A further issue is that both women and men are working in paid employment well beyond the normal retirement age and/or pursuing flexible or partial retirement options. In other words, ageing is having a highly nuanced and diverse impact on people and society, the economy and families. As family composition continues to change in many parts of the world, policymakers will increasingly need to consider the 'beanpole' of multiple generations rather than the extended branches of family trees (Crampton 2009: 35).

Demography and ageing: global and national perspectives

Due credit must be given to the research activity on demography and ageing by various United Nations (UN) departments over a sustained period of time, and particularly the World Health Organization (WHO), with its focus on ageing and health. As the coordinating body on international health, the WHO aims to 'promote health and to keep the world safe and serve the vulnerable' (WHO 2009). Formed in 1948, the WHO has, among many highlights: worked with stakeholders (especially in developing countries) to pioneer the world's first mass vaccination and public hygiene programmes; launched extensive advertising campaigns against risks such as sexually transmitted diseases, AIDS and smoking in pregnancy; and updated, reformulated and redesigned its messages to reflect new public health priorities, such as healthy eating, exercise and banning smoking inside public buildings.

As the WHO states in its most recent fact sheet (www.who.int/news-room/fact-sheets/detail/ageing-and-health):

- Between 2000 and 2050 the number of people aged 60 and over is expected to double, meaning a fifth of people globally will be over 60 years of age.
- By 2050, 80% of older people will be living in low- and middle-income countries.
- Every old person is different: some require care for their everyday needs.
- Some have the level of functioning of a 30-year-old.

The same document suggests the key influences on health in older age groups are:

1. *The individual*: including behaviours, age-related biological changes, genetics and disease.
2. *The environment we live in*: comprising housing, access to transport, social facilities and assistive technologies.

From a public policy perspective, the WHO argues that institutional, organisational and societal attitudes and behaviour will need to transform at a global and national scale in future years to facilitate 'healthy ageing' (see Chapter 2 for a more detailed discussion). This transformation will need to encompass the following changes:

- Changes in how people in society think about and respond to older people and the overall concept of ageing.

- The daily environments that people use and engage with must adapt and develop to become more 'age-friendly'.
- State and private health systems will need to be reconfigured to accommodate the health needs of an ageing population.
- A greater focus will need to be placed on the long-term care needs of an ageing population.

Ageing emerged as a theme in the WHO's priorities comparatively recently on the basis of data generated by the UN Department of Economic and Social Affairs (DESA).

DESA's Population Division collates and leads on demographic issues that other bodies then analyse and extrapolate findings from the data. Its demographic focus is on the total population, its various components (e.g. fertility, mortality, migration) and its increasing urbanising population since 1950. Table 1.1 illustrates the various proportions of people aged 65 and over around the world in 2019. The UN's *World Population Ageing 2019* report (United Nations Department of Economic and Social Affairs 2020) described the world's ageing population as a key demographic 'mega trend', with almost 703 million people aged 65 or over in that year. Table 1.1 highlights significant variations in the proportions of over-65s in the developed, less developed and least developed regions of the world, with the highest percentages found in Eastern and South-Eastern Asia (almost 30% in the case of Japan), followed by Europe and North America. In contrast, Latin America displays a more varied picture, while the smallest proportions (i.e. 2–3%) tend to be in West and Central African nations as well as some Arab States and Afghanistan. Finally, it seems that the populations of many mature, advanced, industrialised nations in Europe and North America are ageing very rapidly (although in the USA the influx of younger migrants has had a counterbalancing effect).

Table 1.1 Selected indicators of population ageing in 2019: proportion of the population aged 65 years and older

Regional distribution	Percentage	Population(million)
Arab States	5	373
Asia and the Pacific	8	4030
Eastern Europe and Central Asia	10	247
Latin America and the Caribbean	9	653
East and South Africa	3	613
West and Central Africa	3	447
Distribution by development status*		
Most developed regions	19	1256
Less developed regions	7	6448
Least developed regions	4	1050
World	9	7715

Note: * Excludes some Pacific island microstates where no data was available.

Selected countries by region	Percentage
Asia and the Pacific	
China	12
Hong Kong	17
Japan	28
West and Central Africa	
Benin	3
Republic of Cameroon	3
Chad	3
East and South Africa	
Angola	3
Mozambique	3
Uganda	2
Zambia	2
Arab States	
Oman	2
Qatar	2
Eastern Europe and Central Asia	
Afghanistan	3
Bulgaria	21
Croatia	21
Estonia	20
Latvia	20
Lithuania	19
Romania	19
Serbia	18
Slovenia	20
Latin America and the Caribbean	
Brazil	9
Martinique	20
Mexico	7
Uruguay	15
Western Europe	
Austria	20
Belgium	19
Denmark	20
Finland	22
Italy	24
Malta	20
Netherlands	20
Portugal	22
Spain	20
Sweden	20
Switzerland	19
United Kingdom	19
North America	
Canada	18
United States	16

Source: Adapted and developed from United Nations Department of Economic and Social Affairs (2020)

Table 1.2 Selected regional indicators of population ageing between 2019 and 2050

	Over-65s in 2019 (millions)	Projected over-65s in 2050 (millions)	2019–50% change
World	702.9	1548.9	120
Sub-Saharan Africa	31.9	101.4	218
North Africa and Western Asia	29.4	95.8	226
Central and Southern Asia	119	328.1	176
Latin America and the Caribbean	56.4	144.6	156
Australia and New Zealand	4.8	8.8	84
Oceania (excluding Australia and New Zealand)	0.5	1.5	190
Eastern and South-Eastern Asia	260.6	572.6	120

Source: United Nations Department of Economic and Social Affairs (2020)

Table 1.2 illustrates the expected growth in the world's ageing population between 2019 and 2050, most notably in less and least developed countries. People over 80 years of age were largely concentrated in Europe and North America (53.9 million) and Eastern and South-Eastern Asia (48.6 million) in 2019. By 2050, we may expect a geographical shift in this age group towards Eastern and South-Eastern Asia (projected 177 million over-80s), with a more modest increase in Europe and North America (projected 109 million over-80s) (United Nations Department of Economic and Social Affairs 2020). However, the UN forecasts even greater growth in the ageing (over-65) populations of Africa, Western Asia, Oceania and Eastern and South-Eastern Asia, perhaps amounting to an aggregate increase of 250% over the course of those 31 years. So, there will be a significant increase in the total *number* of older people, who will comprise a much larger *proportion* of the world's population. In short, ageing will become a more spatially dispersed phenomenon that affects most countries of the world, albeit to varying degrees, as life expectancy continues to increase.

The practical implications of this are shown in Table 1.3, which uses the concept of the old-age support ratio to illustrate that society is shifting towards ever greater reliance on the working population to generate the necessary wealth, public funds through taxation and general workforce to support its ageing population between 2019 and 2050. This ratio is calculated by dividing the number of people aged 25–64 by the number aged over-65, so a low figure indicates growing dependence upon a declining proportion of those traditionally deemed economically active (25–64-year-olds). On a global scale, the old-age support ratio will almost halve between 2019 and 2050. This will impose far greater pressure on taxation taken from those who are economically active, with a very high impact on the labour market for 25–64-year-old employees and more pressure on public health funding and social protection systems, as the risk of illness and disability

Table 1.3 Old-age support ratio: 2019 and 2050 projections

	2019	*2050*
World	5.4	3.1
Most developed regions	2.8	1.8
Least developed regions	6.8	3.5
Least developed countries	10.5	7.1
High-income countries	3.0	1.8
Middle-income countries	6.4	3.1
Low-income countries	10.6	8.2
Africa	10.5	7.6

Source: United Nations Department of Economic and Social and Affairs (2020)

rises with age. Most of the growth in ageing populations in the world's least developed countries will be due to current high levels of population fertility. The least developed nations will see around a 30% growth in their old-age support ratios, which will impose a possibly unsustainable burden on their non-aged populations to generate the wealth they need to develop the necessary infrastructure and services for their ageing populations. The majority of this growth in age dependency will occur in just nine countries: India, Nigeria, the Democratic Republic of Congo, Pakistan, Ethiopia, Tanzania, South Africa, Uganda and Indonesia. It will dramatically transform their population profile from a very youthful pyramid to the much more aged pattern that is evident in many developed countries today. The implications are that unless the economies of these low– and middle-income countries can generate tax revenues or other forms of wealth to support their ageing populations, current concerns over child poverty will be replaced with an age time-bomb relating to how to provide services and support systems for a dramatically different set of social needs.

Ageing as an interdisciplinary research area: its evolution and social science perspectives

To understand how the study of ageing has developed as a distinct area of study, one needs to recognise that most disciplines and subject areas have created their own theories, concepts and paradigms that have evolved through time to make distinctive contributions to ageing research. This section introduces some of these contributions and approaches to illustrate how they have coalesced and helped to co-create ageing as an interdisciplinary area of study, evolving from its roots as a multidisciplinary subject. The intellectual development of an area may pass through various stages as researchers pursue different approaches to create knowledge in response to the research paradigms and problems they seek to address. In the early developmental stages of a subject area such as ageing, the focus is likely to be on disciplinary issues and approaches, progressing through various evolutionary stages as it gathers momentum and then drawing more critical perspectives through multidisciplinary, interdisciplinary and transdisciplinary developments, as illustrated in Figure 1.2.

Mode of inquiry	Symbol	Description
Disciplinarity		Disciplinary knowledge is specific to distinct branches of learning – it has its own procedures, methods, concepts and ways of framing research problems.
Cross-disciplinarity		Cross-disciplinary knowledge is the 'viewing of one discipline from the perspective of another' (Stember, 1991: 4) or the 'importing' of knowledge from other disciplines.
Multi-disciplinarity		Multi-disciplinarity occurs when 'researchers work in parallel or sequentially from [a] disciplinary-specific base to address [a] common problem' (Rosenfield, 1992: 1351).
Inter-disciplinarity		Inter-disciplinarity occurs when 'researchers work jointly but still from [a] disciplinary-specific basis to address [a] common problem' (Rosenfield, 1992: 1351).
Trans-disciplinarity		Trans-disciplinarity is the most collaborative approach to research: 'researchers work jointly using [a] shared conceptual framework drawing together disciplinary-specific theories, concepts, and approaches to address [a] common problem' (Rosenfield, 1992: 1351).
Post-disciplinarity		Post-disciplinarity weaves a unique inquiry thread. It is an 'escape' from disciplines – marked by flexibility, creative problem-solving and intellectual disobedience.

Figure 1.2 Modes of knowledge production

Source: Pernecky (2016: 7)

Numerous disciplines contribute to ageing research, clustering around a common interest that recognises the value of a multidisciplinary approach in the progression and development of knowledge. Shock (1952: 1) outlined the necessity of a broad approach towards the development of 'Gerontology [which] is the scientific study of the phenomenon of ageing … The problems of gerontology are multidimensional and will require for their solution not only a multidisciplinary approach but also a correlation of diverse finding and viewpoints.' As Shock (1952: 1) also indicated, the diversity of research findings arising from multidisciplinarity need to be assessed and reified so as to assist in the development of scientific endeavours around ageing research spanning both science and social science. The broad remit of gerontological research led Shock (1952: 2) to expound the significant contribution of medical research, with its focus on geriatrics as a specialism, interest in the healthcare needs of an ageing population and dual role in preventing and treating diseases and medical conditions. As Hughes (2018) explained, geriatric medicine in the 1960s was primarily concerned with

> immobility, instability, incontinence, and impaired intellect/memory. These 'giants' have changed over the past 50 years. The understanding of 'modern geriatric giants' has evolved to encompass the four new syndromes of

frailty, sarcopenia, the anorexia of ageing, and cognitive impairment. These syndromes are the harbingers of falls, hip fractures, affective disorders and delirium with their associated increase in morbidity and mortality.

Hughes (2018) also explained geriatric medicine's increasing concentration on the'5Ms' – 'mind, mobility, medications, multi-complexity and matters most' – in line with Tinetti *et al.*'s (2017) advocacy of a more patient-centred and holistic approach to healthcare in the ageing population.

In an echo of Cowdry's (1939, 1942) thoughts on the subject of ageing, Shock (1952: 2) suggested that 'The problems of gerontology fall into four major categories: (1) the general biology and physiology of ageing, (2) the psychological changes with age, (3) pathological deviations and disease processes, and (4) the socioeconomic problems of an ageing population.' As Stuart-Hamilton (2006) observed, the middle of the twentieth century witnessed some significant research into the degenerative effects of ageing, leading to two key approaches that developed from psychology and have shaped gerontology in subsequent years:

- *Proximal ageing*, which can be attributed to factors in one's recent past (this has shaped what has been termed 'the life-course approach'); and
- *Distal ageing*, in which conditioning factors in early life have a significant bearing on life chances.

The intellectual roots of gerontology can be traced to influential studies by Hall (1922), as reiterated by Kaplan (1946) and Miles (1942), with their prevailing focus on physical decline in old age, as 'bodily strength, swiftness, and exactness of gross motion tend to fail as the years pass' (Miles 1942: 780). This negative view, and the consequent inevitability of equating ageing with human decline, was an important principle and institutionalised ideology that endured into the 1980s. From a sociological perspective, this institutionalisation, as Townsend (1981) observed, can be dated to the UK's new Poor Law of 1834 and successive governments' attitude towards caring for the country's destitute aged population. It is only recently that new paradigms, such as 'active' and 'healthy' ageing, have come to the fore. These new ideas run contrary to the traditional institutionalised views of an ageing population as passive, dependent and of no economic value to society. Some of that institutional dependency can be traced to nineteenth-century medical research that supported such views. For example, Robert (2006) highlighted Fritz Verzár's early gerontological experiments (conducted in 1886), which demonstrated the decline of muscular strength and cell loss with age. Robert and Labat-Robert's (2017) overview of the history of gerontological medicine highlighted medical researchers' establishment of a link between longevity and nutrition in the nineteenth century as well as the positive impact of vaccination programmes in the early twentieth century, both of which extended life expectancy. Later studies in the field of gerontology, such as *Social Theory and Ageing* (Powell 2006), have not dramatically shifted the focus since Shock's (1952) landmark text, arguing that it comprises a 'science' with biological/physiological dimensions that also considers society's attitudes towards how ageing is constructed and how the aged should behave. Yet, some areas of social science, including sociology, have sought to create a sub-specialism within gerontology (i.e. social gerontology). This 'separates out (1) the phenomena of ageing which are related to man as a member of the social group and society and (2) those phenomena which are relevant to ageing in the nature and function of the social system or society itself' (Tibbits 1960: ix). In an echo of Shock's (1952) approach, Tibbits's (1960) edited collection broadly followed similar strands of intellectual inquiry in identifying four aspects of ageing:

- *Biological* – progressive changes in cellular composition;
- *Psychological* – changes to sensory and perceptual capacity;
- *Situational changes with age* – such as family, community and society-related changes; and
- *Behavioural* – changing self-image and the life cycle.

The nineteen contributors also highlighted the complexity of ageing, differentiating between three key stages in the life cycle:

- Middle age;
- Later maturity; and
- Old age

Overall, Tibbits's (1960) collection introduces many of the issues that continue to dominate social gerontology and policy debates to this day, including:

- The behaviour of older people;
- Ageing and the economy;
- Ageing's impact on labour participation;
- Ageing's impact on state and personal finance in terms of income and pensions;
- Government policy issues; and
- Pressure groups' role in initiating future debates on the structure of civil society (e.g. Edwards 2013), the role of ageing therein and the way in which an ageing population might participate in that society.

Further reviews of the sociology of ageing (e.g. Scheid and Brown 2010) consider retirement as a life event (where the risk of disease and disability increases) that includes interactions with one's family and community. Buraway (2005) saw an important role for 'public sociology' (i.e. its focus on societal problems), whereby social gerontology is designed to improve people's lives, and made a coherent argument for more public engagement between sociologists and a wider range of publics to contribute to solving some of the societal problems posed by ageing. Other notable developments, such as concerns over the mobility of the ageing population on a daily basis (including leisure and daily activities), have emerged in transport research (e.g. Banister and Bowling 2004).

As the contributors to Tibbits's (1960) collection recognised, many key social science disciplines played central roles in the establishment of gerontology. In demography, the concept of the life course has added a more dynamic perspective to ageing research by emphasising the continuity in human development from the cradle to the grave. This adds complexity and depth to explanations of how an individual's development is shaped through time and impacts upon the ageing process, shaping that individual's life chances (Komps and Johansson 2015). The lifespan concept, first outlined by Erikson (1963), built upon the theory of human development in psychology to expand the earlier concept of the life cycle (Rowntree 1901). As Huber (2019) recently explained, the impact of life events at different stages of the human life cycle should provide the focus for more in-depth analysis of ageing. This approach recognises how the environments into which individuals are born and in which they grow up shape and condition their life chances, as well as the educational and economic opportunities afforded to them and the healthcare they receive. All of these factors influence the individual's longevity and the likelihood of living in relative comfort or abject poverty in old age. Bengston *et al.*'s (1997) review of social gerontology and the development of explanations of gerontological phenomena confirmed the dominance of life-course and lifespan theories – a theme which will be further explored in Chapter 2.

As some disciplines developed in-depth gerontological specialisms (e.g. behavioural gerontology), reviews of the progress of research continued to highlight gaps in our knowledge. For instance, some 58 years after the foundation of the *Journal of Gerontology*, Buchanan *et al.*'s (2008) review of 25 years of research found a paucity of studies on behavioural problems among older people. Thankfully, a number of gerontological researchers duly answered the authors' call for more inquiries into cognitive impairment, and especially dementia. A slightly later review by Alley *et al.* (2010) identified stress and coping, the emerging area of successful ageing, health service provision and disability as sociologists' favourite gerontological research topics between 2000 and 2004.

Economists have displayed more interest in ageing and gerontology over recent years (e.g. Schulz 2001; Nyce and Schieber 2005; Clark *et al.* 2004), with Meiners (2014: 63) going so far as to suggest that, 'after decades of being overlooked, the topic "demographic change" has now not only entered politics and economics, but is also giving rise to a frenzy of activity'. Lee and Mason (2010: 151) provided a useful summary of the various positions that economists have adopted towards ageing:

> Some analysts view population aging as economically catastrophic, and others view it as innocuous or advantageous, with most economists located someplace in between. These views of the whole carry over into its parts, such as saving adequacy, pension peril, and intergenerational conflict.

The focus of much economic research has centred around the impact of growing numbers of claimants on state pension systems and private pension schemes, the increased demand that an ageing population places on healthcare systems, health inequalities and the state's role in addressing these issues. Research on the economics of ageing highlights issues about the structure of modern society, particularly how we distribute resources to look after an ageing population. The economics of ageing also highlights the financialisation of ageing by the private sector, including asset release schemes and lifetime mortgages associated with home ownership, which illustrates the challenge of providing for one's own financial sustainability in old age. As Berry (2014) observed, this financialisation has placed a greater onus on the individual, in the UK context, to take personal responsibility for their own long-term financial security. Elements of this can also be seen in other countries. Heinze *et al.* (2011) discussed the spending power of the ageing population and its potential benefits for consumer industries such as retailing, tourism and crafts, while Barclays (2015) and Herrmann (2012) noted the significance of this phenomenon for future economic development.

Geography has developed a sustained interest in ageing research, with several review articles on the geography of ageing (e.g. Warnes 2009; Andrews 2020; Skinner *et al.* 2014) recognising the significance of the field and its connection with 'Gerontology … the academic study of the biological, psychological, and social aspects of ageing' (Andrews 2020: 67). A trajectory of research can be traced to the early development of gerontology as a subject, as 'since the 1950s, a longstanding strand of ageing and health research has been concerned with population ageing. It is an area dominated by the work of demographers and population geographers' (Andrews 2020: 67). Inevitably, geographical gerontology approaches ageing from a spatial perspective. The growing interest in the geographical dimensions of gerontology has emanated from social geography, health geography (and welfare geography) and emergent strands of cultural and social geography in terms of emotional geography, with its focus on human emotions and how these are shaped and respond to the environments people experience (e.g. Davidson and Milligan 2004; Davidson *et al.* 2005). Cutchin (2009: 440) argued that 'the study of the geographical dimensions of ageing has never reached its full potential … only a fraction of the depth and scope of collected theories, concepts and methods of geography have been applied to gerontological thinking and research', although the intersection of human geography and social gerontology has already led to important research on:

- Identity and representation;
- Emotions and embodiment;
- Care and caring;
- Health and well-being;
- Ageing in place and emplacement;
- Living arrangements and environments;
- Urban planning and housing;
- Healthcare services;
- Movement and migration; and
- Demographic ageing.

Another notable development has been the formulation of the concept of therapeutic landscapes, emanating from cultural geography and especially the work of Gesler (1992, 1993). According to Williams (1998: 1193), 'therapeutic landscapes are places, settings, situations, locales and milieus that encompass both the physical and psychological environments associated with treatment or healing, and the maintenance of health and well-being'. This humanistic perspective on ageing intersects with tourism and leisure (e.g. Bell *et al.* 2015), as illustrated by Milligan *et al.*'s (2004) examination of gardening – a leisure activity that offers countless opportunities for spiritual, emotional and physical renewal. In the case of allotments, this occurs in a communal setting and assumes a major role in the lives of many older people. As Milligan *et al.* (2004: 1781) observe, 'such therapeutic landscapes create inclusionary spaces in which older people benefit from gardening activity ... that combats social isolation ... by enhancing [their] quality of life and emotional well-being'. This example intersects with the concept of retirement, which Warnes (2009: 36) described as 'an event, the age at which a person ceases paid work. As a life course stage, the state and period of life of having permanently given up work, usually in old age' (although enforced retirement may also occur due to ill-health, caring for a partner or family member or personal choice). From a leisure perspective, retirement also has a transformational effect on the individual and their family. The conventional definition of leisure as 'non-work time' is reshaped radically when the time obligations of work are replaced with non-work time. Retirement, whether it involves a phased and gradual stepping down from full-time employment or a sudden end to all paid work, almost always has a profound effect on individuals, as will be noted repeatedly in subsequent chapters of this book.

From this brief overview, it is evident that numerous disciplines embraced Shock's (1952) visionary assessment of how gerontological research should progress to build a body of knowledge. More than half a century later, Warnes (2009: 36) succinctly summarised gerontology as 'the study of ageing and old age in all forms of life ... [whereas] Social gerontology focuses on the social circumstances and contexts of human ageing.' Similarly, after acknowledging gerontology's biological/physiological foundations, Powell (2006: 18) suggested that 'social gerontology includes the study of societal attitudes of ageing and how these attitudes influence the perceived needs of older people, both on an individual and structural level'. Social gerontology significantly informs the focus of this book and the intersection and interconnections of ageing with the visitor economy.

This chapter has included a wide-ranging yet necessarily selective review of both the multidisciplinarity that is needed to understand ageing in its societal context and the foundations of gerontological research. As Andrews *et al.* (2007: 151) succinctly pointed out when discussing the contribution of geography to gerontological research:

Geography is clearly one of numerous disciplines which are actively involved in gerontological work that has a geographical element. In some cases (such as sociology, demography and environmental psychology), the spatial perspective has been ongoing and longstanding whereas in others (such as epidemiology, social medicine and public health), place is either a new interest or has been recently (re)discovered, in spite of earlier interests.

This quote highlights the breadth of the multidisciplinary research effort that has developed around ageing and place, and it has particular value in capturing the settings in which the broader visitor economy exists. The continued broadening of research interest in ageing through multidisciplinary and interdisciplinary endeavour is reflected in both the focus of this book and its argument that the visitor economy needs to embrace a greater understanding of ageing and demography if it is to adapt to and recognise the global challenges and opportunities ageing populations create.

This chapter has illustrated many of the influences, paradigms and disciplines that have contributed to the study of ageing in order to contextualise the book, identifying the contributions that have been made to the study of gerontology. The thinking that is implicit in these disciplines offers many interesting and novel concepts that we will explore throughout the remainder of the book to explain how ageing-related issues in the visitor economy can be better understood from a variety of disciplinary perspectives. The scale and scope of the ageing research agenda has expanded exponentially since the 1950s, especially its scientific analysis in the field of gerontology, which provides a focal point for this book in relation to the visitor economy.

Structure of the book

Chapter 2 poses three critical questions. How is ageing conceptualised and understood? What is its link to the visitor economy? What research has been undertaken in this area? Key influences on the ageing leisure participant are examined together with economic challenges relating to ageing, including the concept of the 'haves' and 'have-nots', financial resources, active ageing, multi-generational families and the changing structure of living environments as well as the opportunities that seasonality presents for the visitor economy with ageing markets.

Chapter 3 examines ageing as a societal challenge, framing it around accessibility and concepts that inform thinking on these issues, including civil society, universal design, well-being, the role of technology and state intervention. Chapters 4 and 5 explore the accessibility challenges faced by visitors with sensory issues, learning difficulties and/or mental health, physical and degenerative conditions. Chapter 6 examines the visitor economy's current adaptation strategies to accommodate these travellers' needs and suggests that the sector should pay more attention to change management, best practice and communities of practice. Finally, Chapter 7 provides a synthesis of the research themes covered in the book, recommends some avenues for future research and, importantly, offers suggestions as to how the visitor economy might be made more age friendly.

2 Ageing, the visitor economy and a leisure society

Introduction

This chapter seeks to encapsulate the expansion of research on ageing as outlined in Chapter 1 by examining the interactions between ageing and the visitor economy as an approach that connects leisure, recreation and tourism in one continuum with the individual life course and experience of leisure in older age. Academic research on ageing can be located along a continuum that broadly describes two interconnected paradigms. At one end of this continuum is a more pessimistic – and perhaps more traditional – biologically and physiologically informed view that posits that an ageing body, as a stage in the life course characterised by change (Harper 2006), poses particular challenges for individuals. As Nair (2005) suggests, ageing bodies are confronted with substantial hazards associated with the use of space and the environment; for example, the ageing population face a greater risk of trips, slips, falls, fire, hospitalisation, impairments, and risks associated with driving, travel and flying, and a proportion of the elderly are confined to residential care. Juxtaposed with this somewhat sobering assessment are the more enabling, positive and less barrier-restrictive analyses that view ageing as a liberating, new phase of life which is full of potential. Such analyses challenge the traditional association of ageing with inevitable decline and morbidity, and seek instead to promote a more *positive notion of ageing*. To substantiate these arguments, alternative constructions of health and ageing have been formulated that embrace the concepts of well-being and quality of life as essential elements of a positive experience of ageing. New concepts, such as 'successful ageing', stress 'intervention aimed at supporting older people to avoid disease and disability, maintain high mental and physical functioning, and remain socially engaged' (Stephens and Breheny 2018: 3) – a development that Minkler and Fadem (2002) also advocated. Other conceptualisations, such as 'active ageing' (e.g. Walker and Aspalter 2015), articulated a more humanistic and less deterministic/fatalistic construction of ageing to support living well in later life. The World Health Organization (WHO 2007b) has extended this notion of positive ageing in its promotion of 'age-friendly cities' (a subject to which we will return later in this chapter). The approach and imagery which organisations in the visitor economy use in relation to ageing, whether it has a positive or negative connation, will shape

DOI: 10.4324/9781003039358-2

the type of experiences provided. These experiences are usually consumed within one's leisure time, of which tourism is one component (Hall and Page 1999).

In a leisure context, in which tourism and the visitor economy are both located, approaches to ageing have a long history of academic analysis that is often overlooked (e.g. Burns 1932; Kaplan 1960; Lundberg *et al.* 1934). Burns (1932: 183) appositely commented that 'The proportion of the population, in any country, of different age at different dates makes a great difference to the uses of leisure.' Burns defined leisure as follows:

> spare time or leisure must be taken in its most inclusive sense. It is understood here to include all that part of life which is not occupied in working for a living. Energy is expended and vitality grows in leisure; but in work time the direction of energy is controlled by public need or the desire for gain, whereas in leisure we are free. For a very small proportion of adult men and women, leisure is the whole of their lives. They belong to what is called the 'leisured class'. Some of them have retired from active work with enough income saved; and a small proportion of this 'leisured class' has inherited the power to live in leisure without having done any work for a living.
>
> (Burns 1932: 19)

Burns (1932), then, created the basis upon which many future definitions of leisure would be formulated (see Page and Connell 2010). Despite this, almost half a century later, Glyptis (1981: 311), reflecting on debates about the definition of leisure, argued that leisure is related to activities, attitudes, behaviour and individuals' characteristics, concluding that:

> The very concept of leisure is elusive, and permits of no single, simple definition. It is usually taken to refer to discretionary time or activity (Countryside Recreation Research Advisory Group 1970), or the attitudes surrounding them (Parker 1976), and by any measure involves a large proportion and diverse range of human behaviour.

With regard to characteristics, Burns (1932) examined early-twentieth-century innovations in leisure provision for those in 'old age' (i.e. those aged between 60 and 80), with particular reference to motor cars, cinemas and aeroplanes, all of which would continue to provide a multitude of leisure opportunities for future generations (e.g. tourist travel) alongside a growing commercialised and mass leisure economy (Page and Connell 2010; Beaven 2005; Marc and Meyersohn 1955; Moore and Van Nierop 2006; Surdam 2015). Burns (1932: 25–6) went on to explain why so many older people could afford to take advantage of these innovations:

> The increase of incomes among manual workers and low-paid salary earners is due … to improved benefits from insurance against ill-health, accident, or old age, to new pension-funds and to better facilities for 'savings' … The expenditure on leisure, therefore, of those with small incomes has increased, partly because a greater number of 'bare needs' are supplied 'in kind' from public funds.

Burns here represents the positive approach to ageing, which highlights the fact that older people have a greater disposable income for leisure (including pensions); this is epitomised in Kleiber and Genoe's (2012) use of social theory to illustrate the positive benefits of leisure in later life. Research has significantly progressed since Kaplan (1960: 408) argued that 'the expectations and interpretation of leisure for the retired person are vague'. Alongside Burns's (1932) study, a range of surveys of communities (e.g. Jones 1934; Rowntree and Lavers 1951), constructed in the social survey tradition, placed emphasis upon issues such as ageing and the social condition of the population per se.

The negative and positive positions on ageing will resurface throughout this book as a tension between the desire to recognise the limitations that ageing may present to some people and the desire amongst others to remove barriers to engaging and participating in the visitor economy. The book's main thesis is that the visitor economy offers many opportunities for more fulfilling ageing leisure lifestyles. Glyptis (1981: 314) defined 'leisure lifestyles' as follows:

> the aggregate pattern of day-to-day activities which make up an individual's way of life is here defined as his [*sic*] life-style. This encompasses both work and leisure … a life-style approach involves three important shifts of emphasis from previous participation studies: from activities to people; from social aggregates to individuals; and from expressed activities to the function which they fulfil for the participant and the social and locational circumstances in which he [*sic*] undertakes them.

This chapter considers one overarching question: how should we conceptualise and understand ageing? As highlighted in Chapter 1, approaching this question in a meaningful way required perspectives to be drawn from a range of contributing disciplines and subject areas, encompassing sociology, leisure studies, tourism, economics, marketing, politics and health science to construct a broad conceptualisation of ageing and its interconnections with the visitor economy. While this conceptualisation is necessarily selective and concise due to the constraints of space, it does provide a distinctive focus for a more holistic understanding of the ageing–visitor economy nexus. First, though, we must consider what we mean by the term 'visitor economy'.

The visitor economy: a new concept for analysis?

The term 'visitor economy' is not widely used in the academic literature (the exception being Smith and Graham 2019); rather, it is found most frequently in public sector settings. Law (2002) used the term when describing its use by political decision-makers to broaden out the remit and scope of tourism and associated areas. This is illustrated well by Tourism Toronto, for whom 'the visitor economy … [is] a term much broader than tourism. Toronto's visitor economy encompasses the direct visitor spending in the destination and the indirect and induced economy activity that stems as a result' (Tourism Toronto Partners 2019). Box 2.1 provides global examples of countries, destinations and organisations that have adopted the visitor economy as a focal point. The purpose of the visitor economy concept is

Box 2.1   Examples of tourism and associated bodies using the visitor economy concept in strategies, reports and briefing reports

- Local Government Association, UK
- Liverpool Local Economic Partnership, UK
- Transport for the North, UK (*https://transportforthenorth.com*)
- North West Regional Development Agency/Culture North West, UK
- Bath and North East Somerset Council, UK
- Parliament of Scotland/VisitScotland, UK
- Visit Brighton, UK
- West Midlands Growth Company, UK
- Visit East of England, UK
- Gloucestershire First Local Economic Partnership, UK
- Horsham District Council, UK
- Greater Lincolnshire Local Economic Partnership, UK
- Foundation for Puerto Rico
- Welcome to Wales/Welsh Government, UK
- Visit Britain, UK
- Destination London, UK
- Cumbria Local Economic Partnership, UK
- South Australian Tourism Commission, Australia
- Tourism Industry Council South Australia, Australia
- Tourism New Zealand, New Zealand
- Destination Auckland, New Zealand
- Northern Ireland Tourist Board, UK
- Victorian Government, Australia
- Vienna Tourist Board, Austria
- North Sydney, New South Wales, Australia
- Gladstone Region, Queensland, Australia
- Western Sydney, New South Wales, Australia
- Tasmania, Australia
- People 1st, UK
- Isle of Man
- European Cities Marketing Association
- Brisbane, Queensland, Australia
- Bay of Plenty, New Zealand
- Greater Manchester, UK

to demonstrate much greater economic benefits from visitor activity than would be measured through tourism alone, to draw upon the wider interconnected visitor economy as has been modelled through initiatives such as Tourism Satellite Accounts (Page and Connell 2020). Deloitte and Oxford Economics (2008: 17) adopted a narrow conceptualisation that included two components (direct and indirect effects):

- A core tourism concept that focuses on the direct contribution of tourism activities (i.e. the value added generated by the provision of tourism-characteristic goods and services). This measure is in line with the concept used by the United Nations World Tourism Organisation (WTO) Tourism Satellite Accounts (TSA).
- A broader measure that also takes into account indirect effects (via the supply chain) as well as the impact of capital investment and collective government expenditure on behalf of the tourism industry (the total Visitor Economy).

'Estimates of the economic contribution of [sic] visitor economy in the UK are then defined using both value-added (VA) and employment … in addition a key measure used by VisitBritain and other government agencies … is the total spend by visitors–the size of the tourism market–including both inbound and domestic tourism' (Deloitte and Oxford Economics (2008: 17).

This approach emanated from a desire to measure the economic value of the visitor economy. It was subsequently reformulated by Connell *et al.* (2017: 111), building upon a report by Barclays (2015), to develop the following definition that is better suited to the purposes of this book:

It embraces the hospitality and tourism sector (food and drink provision via cafes, restaurants and accommodation), travel agencies, transport providers, cultural activities like galleries, events and retailing. There is often a blurring of the terms visitor economy, tourism and leisure as residents may also use the facilities and services in their leisure time. The term broadly refers to the supporting infrastructure that caters for the needs of visitors and residents especially in their leisure time and so is very wide ranging in what is included in such a categorisation.

This definition transcends the traditional tourism–leisure nexus inherent in continuums that incorporate tourism as a function of leisure (Hall and Page 1999; see Figure 2.1), providing a more comprehensive and contemporary concept that recognises the types of services that people use in their leisure time. Moreover, the

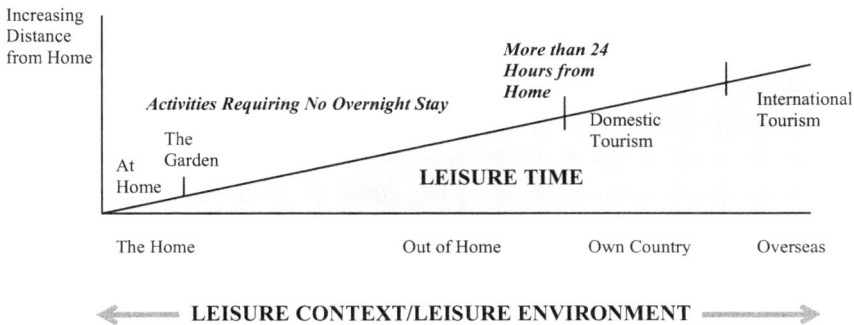

Figure 2.1 The leisure continuum

Source: Developed from Page and Connell (2010)

notion of the visitor economy as a broad economic sector has particular salience as the focus of this book is older people's multiple interactions, or touchpoints, with different businesses and settings. These touchpoints were defined by Connell and Page (2019b: 32) recognising that

> The complexity of destinations, the number of elements that interact as part of the visitor experience and the large number of touchpoints in the visitor journey, create a significant challenge ... One of the key challenges for any destination is to understand the nature of these touchpoints and where they take place. Furthermore, it is crucial to know where these touchpoints have the greatest significance for people ... and where the greatest opportunities and challenges exist.

(See Figure 2.2 as an example)

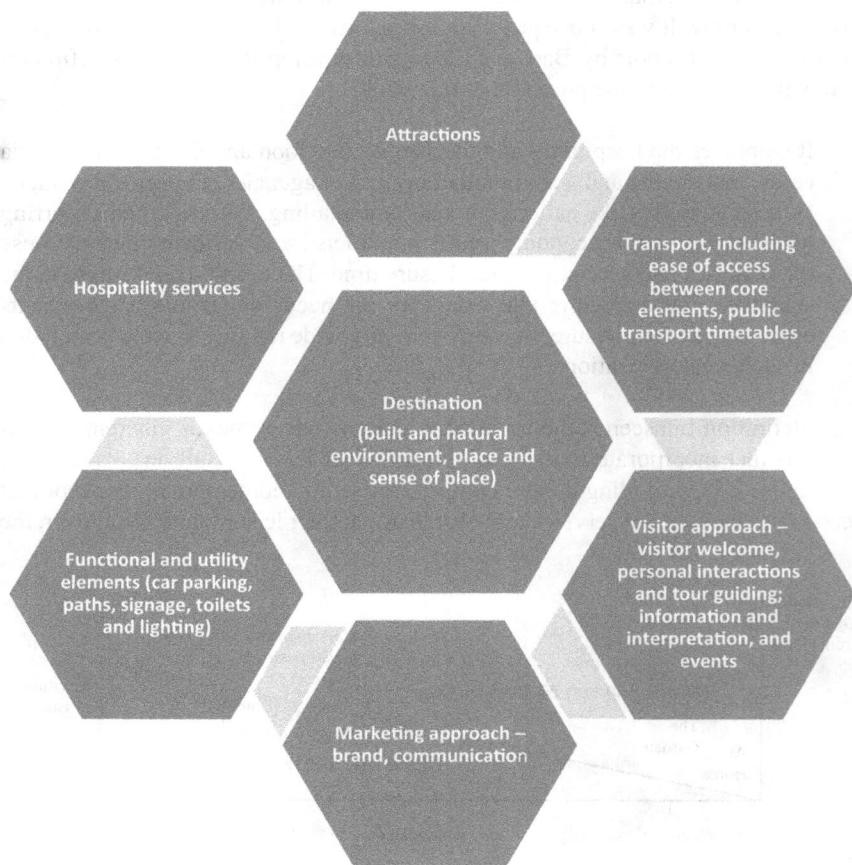

Figure 2.2 Touchpoints in a destination and the visitor economy

Source: Reprinted from J. Connell and S. J. Page, 'Case study: Destination readiness for dementia-friendly visitor experiences: A scoping study', Tourism Management, 70, p. 33, © (2019), with permission from Elsevier

Conceptualising ageing: what is old age?

'Old age' is a challenging term to explain, and one not without controversy in its application and interpretation. Some of the complexity is illustrated in Box 2.2, where a wide range of terms and criteria are used to identify the point at which old age begins in the human life cycle. Defining old age can be problematic because:

- First, it has traditionally been defined as a chronologically fixed point in time, with no reference to individuals who may not perceive themselves to be 'old'. Law and Warnes (1976: 453) highlighted the challenge of using such chronological definitions for retirement migration where

> no single definition of the elderly is satisfactory for all purposes. We are particularly interested in the group who have retired from full-time employment, which is a variable age for individuals and may be anything from 55 upwards. Whilst there are some statistics for the economically inactive, it is not always possible to use them to define the retired, mainly because of the ambiguous status of women. Many studies have therefore relied on age rather than economic status, and used the pensionable ages of 65 for men and 60 for women, to define the elderly or retired [prior to the harmonisation of retirement ages of men and women in the UK].

What this discussion illustrates is that 'retired' does not necessarily equate to 'old age'.

- Second, the increase in life expectancy has called into question the traditional, chronologically determined criteria for when old age begins. Roebuck (1979) illustrated this in the UK: the 1875 Friendly Societies Act adopted a definition of old age as 'any age after 50'. The WHO currently considers the period after the age of 60 as 'old age', whereas most developed countries use the age of 65. The latter chronological definition is typically when state or private pensions (i.e. the Normal Pension Age) become payable. But one's biological age and calendar age are not necessarily the same thing. As the WHO (2002) highlighted, in Africa, for example, the question of whether it would be feasible to adopt an age range of 60–65-plus would depend upon the socio-economic conditions in the different nations (e.g. life expectancy, health conditions and living standards). A further factor in Africa was that actual birth dates were often unknown (WHO 2002; Togunu-Bickersteth 1987, 1988). Such issues underline that while old age is biologically inevitable, how it is defined varies according to the ways in which different societies socially construct and perceive it (see Johnson 1976; Brubaker and Powers 1976). One facet of socially constructed meanings of old age relates to the roles which older people play in society, especially the physical and psychological decline that reduces their ability to participate in particular activities, such as work. The result is that old age has been seen as synonymous with retirement. Other components of this debate include the point at which 'older age' sets in and when mental acuity and other capabilities start to diminish (Thane 1989). Organisations such as the British Geriatric Society have illustrated the importance of debating definitions and usage of 'old age' in view of the need to rethink how we conceptualise the term.

- Third, no two human beings are the same, so what may well apply to one person as a categorisation may not apply to another. Ageing is affected by a number of factors including genetics, career, lifestyle and attitudes. Different generations have different life experiences that affect perceptions and attitudes to life and ageing, which is important if one adopts the broad categorisation of the ageing population as being over 50 years. This grouping encompasses two or more generations if one adopts the definition of a generation as being twenty to thirty years (Berger 1960), or based on the spirit and thinking of a group of people through time (e.g. the Silent Generation, born 1928–45, is radically different to Generation X, born 1965–80). Furthermore, later marriage, multiple marriage and later parenting means that some people are becoming parents in their 40s and 50s, sometimes for the first time (see Ylänne 2016), which has significant implications for leisure time and experience. In addition, as Freund (2020: 499) argues, the intensity of a person's life through middle age, including the pursuit of career goals and having a family may result in the deferment of leisure until retirement, something that Freund terms as the 'bucket list effect'.

Box 2.2   Definitions and classifications of older adults: key issues

- According to Patterson (2018), a wide variety of terms, including 'baby boomers', 'the senior market', 'the mature market', 'the grey market', 'young singles', 'young senior generation' and 'woopies' (i.e. well-off older people), are used to define older adults, often interchangeably.
- Different classifications have been used by marketers to segment the market of ageing populations, including 'the silent generation' (born between 1928 and 1945). Patterson (2018) suggests that this generation has been characterised as travelling infrequently for tourism.
- Schiffman and Sherman (1991) described the 'new age elderly' as more adventurous, pursuit-oriented and interested in challenging activities.

They are characterised as around 12 years below their chronological age in terms of behaviour. Their independent, travel-oriented lives set them in sharp contrast to the silent generation.

- 'Seniors' is often used to describe people over 55 years of age. They are frequently characterised as having a strong affinity with tourism and leisure travel.
- A term used in generational marketing, 'baby boomers' (born between 1946 and 1964; Page and Connell 2020) comprises a generation that has enjoyed the economic and political stability and wealth of the post-war period. While the members of this group may not always consider themselves 'elderly', they have enjoyed several unifying features in Western economies: affluence, increased longevity, fewer children, and travel diversity from owning cars and travelling domestically and internationally.

(Developed from Patterson 2018)

More holistic demographic and sociological analyses of how to construct 'old age' suggest that chronologically determined ideologies of age need to be replaced with more flexible and fluid approaches. Sanderson and Scherbov (2015) looked at three existing approaches to ageing, namely:

- Age structure changes and reasons for these changes (e.g. fertility and mortality) and the traditional measure of those aged over 65;
- The age-specific characteristics of the population (i.e. how many people over 65 will be in the population); and
- The changes in ageing and the relationships and interactions between age structure and the age-specific characteristics of the population in the future.

Drawing upon data from 39 European countries, they then suggested we need to shift towards a new threshold for old age that is far more dynamic and fluid by focusing on the number of years of life expectancy that remain (e.g. 10, 15 or 20 years).

Swift and Steeden (2020: 5) also challenged the calendar-years approach on the grounds that 'our understanding of age and ageing is socially constructed and encapsulated in commonly-held age stereotypes, internal representations of what younger and older people are and should be'. The European Social Survey for 2007–2008, cited by Swift and Steeden (2020), pointed to significant variations by country, with an average of 61 years accepted as the point at which old age begins (see Figure 2.3). Women, compared to men, perceived the point at which old age begins as coming later. Swift and Steeden (2020) also cited data from a 2018 IPSOS study in which respondents aged 16–64 perceived old age to commence at age 68, on average. Therefore, it is evident that conventional measures and definitions of old age need to be viewed as fluid and evolving beyond the established thresholds of 60–65 years of age in relation to the attitudes and views of society.

To illustrate just how difficult it is to define 'ageing', we asked our respondents to provide immediate responses to the term. Those presented in Box 2.3 comprise a range of very grounded and perceptive insights from practitioners who deal with issues of ageing on a day-to-day basis and recognise some of the complexities surrounding creating a working definition. This not only reflects the fluidity with which we need to approach ageing as a phenomenon but reinforces the previous discussion about conventional stereotypes and the tendency of the media and other groups to conflate a range of age groups and label all of them 'old' (see Yianne 2016). In addition, the quotations that follow illustrate the nuanced attitudes of different age groups towards the term 'ageing' and suggest that different life chances might be associated with participation in society. Finally, it should be noted that Respondent 15 concluded that we live in a fundamentally ageist society – a point to which this discussion will now turn.

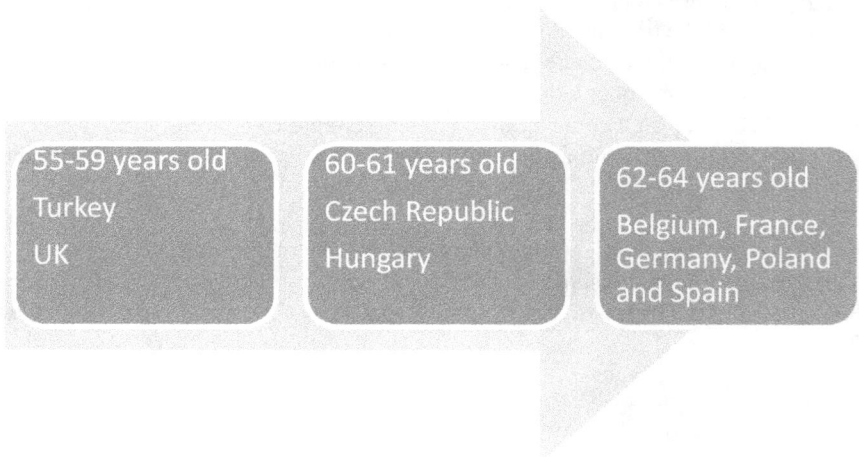

55-59 years old	60-61 years old	62-64 years old
Turkey	Czech Republic	Belgium, France, Germany, Poland and Spain
UK	Hungary	

Figure 2.3 Age at which people are considered 'old' in various European countries
Source: Adapted from Swift and Steeden (2020), based on the European Social Survey

Box 2.3   Responses to the term 'ageing'

We're ageing from the day that we're born. It's a fact of life that we age, so it isn't a problem and it's not a disease, it's only a fact of life.

(Respondent 1)

Well, it's often a pejorative term but it's the only certain thing in life in the same way that death is. So you ignore it at your peril. It usually sneaks up on you very quickly. You still have a mindset at my age, because in two weeks' time I'm 65, I don't know what the hell happened there. How did that happen? Where did it come from? So you never regard yourself as being part of it unless you're quite honest with yourself about ageing. So it's something that applies to your parents if they're still alive or other people's parents or other people. You tend not to look at yourself unless, you know, you do have an acceptance of where you stand in the scheme of things.

(Respondent 2)

I'd have to put it within the context of my work because each year we look after around 15,000 people and the vast majority of them are old and in fact the vast majority of them are older old, so they're sort of 80-plus really. So my understanding of the phrase 'ageing' is very much sort of influenced by that and I would say that it can be a really positive process, it can be a really good time when you feel part of your community, you feel supported, but in reality it's often not that, it's often a time of increasing constriction on life, being unable to leave home, resenting support but needing support, probably not accepting support to remain independent soon enough and as a consequence requiring more support for your needs.

(Respondent 3)

So, my perception of ageing, gosh, that's a very open question. So, I think my perception of ageing is probably informed from both my work and my personal experience. So, from … personal experience, I consider myself lucky that – so far – I am ageing, and not dead. And from a business perspective, I think ageing is actually a natural progression of human life that perhaps has a less negative … connotation in the Western world. I think 'ageing' as a term is not seen as positive, whereas actually, if you're lucky enough to be ageing, you should be celebrating.

(Respondent 5)

First impression, it makes me think of England because you spell it with an 'e' … I've worked with a not-profit that serves older adults, so I have a very positive association with the term.

(Respondent 6)

My view of ageing is that from about the age, I mean it varies, but from about the age of 70 to 75 onwards we see a deterioration in many respects of people's circumstances in terms of physical mobility, sometimes cognitive decline, but more importantly a diminishing of social support because friends and family can be more limited as people grow older as they often outlive their friends and family if they have a very small family network. I think there's a tale of two sides of ageing. There's the one that we won't see … because the people that have large support networks, or do not suffer from the effects of isolation, or are physically very mobile and independent, you know, can be [fit] until their nineties [but then again], we see the people that are the other side of the spectrum who have health problems, whether they be physical or mental, and have issues around loneliness and isolation. They're the people that we come into contact [with]. So the term … 'ageing' really evokes from me an array of different perceptions about age.

(Respondent 10)

[I]t's a process that affects everybody … Effectively, if you're defining people by age, really you need to be starting within the employment angle of the 50-plus. But I think that it's important to understand a minimum of three different demographics within the over-50s. The older working age, 50 to retirement age, 66. The active pensioners, sort of 66 to 80, who are … the mainstay of volunteering and quite often [do] a lot of caring. But also, of course, there are about 1.2 million of them [in the UK who] are still in work, even if it's part-time. And then it's only really the over-80s who start to exhibit a majority of people who've got disabilities, health conditions and mobility or cognitive issues. But, unfortunately, it appears to me that a large number of decision-makers in the media and the public conflate the over-50s with the over-80s, which would be, in the same way, to conflate the 50-year-olds with the 20-year-olds, which makes no … sense. So, I think that we need to be very clear about what we are talking about when we're looking at age. And I also think that we live in a very fundamentally ageist society and economy.

(Respondent 15)

Ageing and society: issues associated with ageism

Age prejudice is a human rights violation, exhibited in areas such as healthcare provision and employment (Butler 2006). It is embodied in the concept of ageism, which is defined 'as a process of systematic stereotyping and discrimination against people because they are old ... it is deeply ingrained in society, categorising old people as senile, rigid in thought and manner, and old fashioned in morality and skills' (Butler 2006: 41), reflecting the negative attitudes society often has towards older people. The Irish Longitudinal Study (Gray and Dowds 2010) reported very positive attitudes towards older people between 2003 and 2008. Other nations and groups positively associate age with wisdom and respect (e.g. Greece, Native Americans, South Korea, China, India, Japan, Vietnam and certain tribal societies). Nevertheless, ageism is a background issue in many countries as debates about poverty, child neglect, climate change and economic cycles (i.e. recessions) attract greater media and public attention.

Kocchar and Oates's (2014) global study of attitudes towards ageing confirmed this:

- In response to the question 'Is ageing a concern in your country', 87% of Japanese respondents stated that it was a problem. (At the other end of the scale, just 23% of Egyptians agreed that it was an issue.)
- In most countries, responsibility for the economic well-being of the ageing population lies with the government and individuals.
- Older people think old age is a major problem in Israel, Britain and Brazil.
- With regard to people having an adequate standard of living in retirement, 70% of respondents in China, Brazil and South Africa were 'confident', compared with only 20% of Russians and Italians.

The WHO, referring to their analysis of the World Values Survey (2016), found the lowest levels of respect for ageing people occur in the high-income countries, reflecting the ingrained nature of ageism. This conclusion is supported in reports published by organisations that have an advocacy role for an ageing population (e.g. Age UK and the Centre for Ageing Better in the UK). For example, Hill (2020) reported that 'older people [are] widely demonised in [the] UK' in an article that drew inspiration from the Centre for Ageing Better's (2020) *Doddery but Dear? Examining Age-Related Stereotypes*. This report indicated that:

- Older people are viewed as wiser but less visible;
- Older people are seen as possessing warmth but low levels of competence, so they attract pity and are patronised;
- Societal attitudes to older people are sometimes characterised as benign indifference;
- European Social Survey data found that 41% of respondents were disrespected due to their age;
- Older workers are viewed as reliable and dependable, but also weaker in terms of overall performance, less willing to learn and more expensive than their younger counterparts;
- Old age is perceived as a process associated with poor health, a connection which is sometimes exaggerated; and

- The media stereotype older people as a burden through use of metaphors such as 'grey tsunami', 'demographic cliff' and 'demographic time-bomb' but also glamorise images of an ageing lifestyle that is out of reach of most people.
(Developed from the Centre for Ageing Better 2020)

The Centre for Ageing Better presented ageing as a macro issue for governments, policymakers and the general population, and more specifically younger age groups, whose perception tends to be generally negative.

Having examined the varying societal attitudes towards ageing, we now turn to how we might interconnect the principal themes of this book – ageing, tourism and the visitor economy – from both a conceptual and a theoretical perspective.

Conceptualising and theorising the ageing – visitor economy relationship

Conceptualising and theorising in social science perform an important role in helping to make sense of the world around us by simplifying the complexity of reality and seeking to help us understand how human and natural phenomena operate, exist and develop using general principles and rules that help to explain the reality we observe. Theory is often an abstract concept and outlined in a seemingly erudite manner by many scholars, typically during communication with other scholars in the same discipline who are familiar with the language and principles used to question ideas and interpretations of the phenomenon they seek to explain. It is often used to build an underpinning framework to guide the research questions that are framed and examined during the research process. We also find that 'the history of science reveals that theories have nearly always predated the availability of appropriate methods for testing them' (Godbey *et al.* 2010: 129), so theoretical developments set the context for future empirical assessment of their validity.

Alley *et al.* (2010) reflected on the 65 years that had passed since the foundation of the *Journal of Gerontology* and the use of theory in ageing research. They succinctly pointed out that theory is used 'to explain why phenomena occur', so it is an important process in increasing our understanding of ageing (Alley *et al.* 2010: 583). Social gerontology, informed by its constituent discipline – sociology – provides an important starting point in the pursuit of understanding ageing from a theoretical standpoint, although other disciplines have also contributed to the development of this specialist area of academic endeavour. As Alley *et al.* (2010: 583) pithily concluded:

> Theories of ageing help to systematise what is known and explain the how and why behind the what of our data … Theory serves at least three critical purposes in research on ageing: to guide research questions and hypotheses, to help explain research findings, and to inform interventions to solve ageing-related problems.

Other reviews of social gerontology (e.g. Bengston *et al.* 1997) have comprehensively surveyed the theories used to study ageing, notably some of those introduced in the previous chapter (e.g. the life course and the lifespan; see Harper 2006) as well as many other theoretical perspectives. The latter include phenomenology, critical theory and perspectives that span sociology and psychology with a single unifying feature – their application to ageing. In a similar vein, Alley *et al.* (2010) outlined a range of theoretical and conceptual models in the social gerontological literature to illustrate the depth and specialised nature of the research conducted on different aspects of ageing. Given our focus on the visitor economy, we are most interested in the evolution of theoretical perspectives that are associated with the expansion of leisure time in later life and its consumption, sometimes associated with leaving the labour force (i.e. retirement). As Argyle (2001: 107) argued, 'on average the retired are more happy than those at work'. However, this needed some qualification: those with stressful or boring jobs were pleased to retire whereas those with more interesting jobs were less willing to do so. Argyle (2001: 108) proceeded to argue that 'retirement has often been classed as a "stressful event" ... and it does take people a few months to get used to it'. Citing Stockdale (1987), Argyle's (1992) earlier examination of retirement as part of everyday life found leisure to be an important aspect of older people's lives. Yet, with increasing longevity, the relationship between ageing, work and retirement has started to blur, which led Argyle (2001) to stress that some people retire due to ill-health while others choose to retire either earlier or later than the state pension age. This seemingly complex situation may be best described as the 'leisure paradox'. This is worthy of further exploration before we turn our attention to the social theory that has been developed to examine leisure and ageing, especially the relationship between ageing, leisure and retirement.

The leisure paradox in later life

Many studies in the 1960s categorised Western nations as emergent *leisure societies* in which leisure time was increasing (e.g. Dumazedier 1967). This research built on Veblen's theory that leisure time diffused down through the social classes (Veblen 1953; originally published 1899) following a process of emulation. As we saw earlier, a number of subsequent studies questioned how the working classes would use this new-found leisure time, whereas others adopted a counterposition, arguing that the quality of leisure was declining amongst a 'harried' working population (e.g. Linder 1969). This latter thesis suggested that the working population was suffering from leisure time compression as they tried to cram too many things into increasingly busy lives. It was subsequently revisited by Schor (1993), who openly challenged Veblen's thesis. Other studies, such as Young and Willmott (1973), pointed to the dual burden of paid and unpaid work since the 1960s, which had reduced women's leisure time.

These broader changes help to contextualise an expanding ageing population that poses a challenge for how we characterise the 'leisure paradox'. Figure 2.4 presents a highly generalised illustration of this paradox, with leisure simply defined as non-work, even though we recognise that there are other, more complex definitions of the term. For instance, after stating that American men gained an extra 5.26 hours a week for 'leisure' in the period 1965–2003, Sevilla

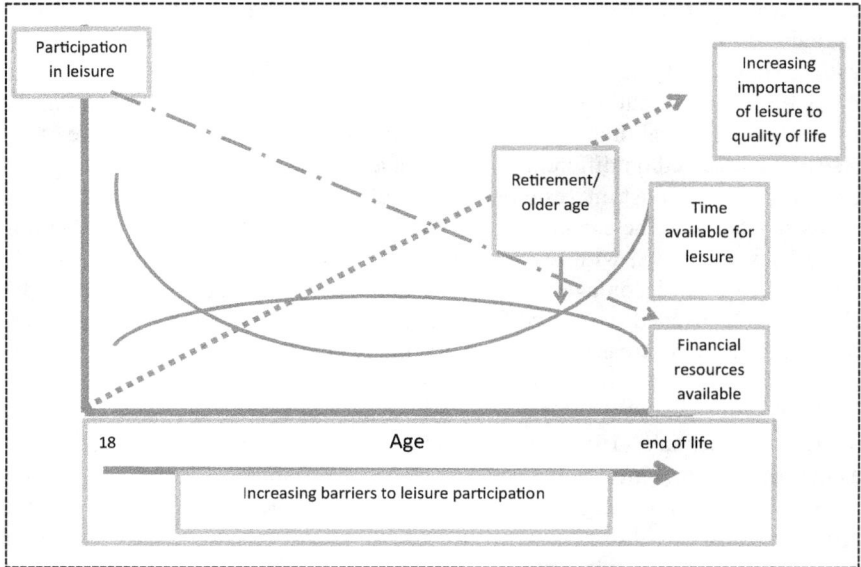

Figure 2.4 The leisure paradox

et al. (2012) suggested that this should be subdivided into *pure* leisure (i.e. traditional leisure activities), *co-present* leisure (i.e. social activities) and leisure *fragmentation* (i.e. declining opportunities for leisure due to competition with other activities).

Figure 2.4 suggests that the leisure paradox emerges in later life, primarily because leisure becomes more important for the individual's quality of life and assumes greater meaning at that age, yet, paradoxically, ever more barriers increasingly restrict participation (see also Nimrod and Shrira 2016). These barriers may be social-psychological, cognitive, cultural-environmental or technical (e.g. mobility). Of course, there are some obvious exceptions to these general principles: for instance, some people undertake part-time or volunteer work after retiring from their main careers (so-called 'encore careers'). That said, the leisure paradox certainly demands further attention. With that in mind, we searched through a number of data sources for information on available income and time for leisure in relation to age. Table 2.1 presents the mean weekly earnings of full-time workers in the United States and shows that earnings gradually increase over the course of a typical working life to peak in age group 55–64 for men and age group 35–64 for women before declining significantly thereafter. Men's incomes are considerably higher than women's in all age cohorts. Turning to Table 2.2, which is based on the American Time Use Survey, leisure time drops after the age of 24 and does not rise again to a significant extent until the

Table 2.1 Median weekly earnings of full-time wage and salaried workers by age in the United States, 2018 (fourth quarter)

Age	Male (US$)	Female (US$)
16–19	513	451
20–24	651	585
25–34	917	815
35–44	1162	923
45–54	1174	909
55–64	1188	922
65 and over	1099	822

Source: Adapted from US Bureau of Labor Statistics, www.bls.gov/

age of 65; conversely, time spent on work rises significantly after the age of 24. To some extent, the latter figure is explained by the amount of time that is spent in education up to the age of 24. In the UK, Table 2.3 shows that median gross earnings increase up to the age of 49 and then start to drop from age 50. Other studies, such as the UK's Time Use Survey and national surveys, corroborate the general increase in leisure time in later life. For example, the Office for National Statistics' *Leisure Time in the UK* report (ONS 2015), which was based on UK Harmonised European Time Use Survey data, found that average leisure time ranged from 6 hours per day in the 16–24 age group to 4.8 hours for those aged 25–44, 5.3 hours for those aged 45–54, 6 hours for those aged 55–64 and 7.2 hours for those aged 65 and over. Differences between men and women were consistent with those observed in the United States, where men spent an average of 6 hours 9 minutes a week on leisure, compared to 5 hours 29 minutes for women (typically due to the fact that women do more unpaid work than men). Among the over-65 age group, men had 45 minutes more leisure time each day than women (see Small 2003; Sedgley *et al.* 2006). Therefore, as Nimrod and Shrira (2016) point out, ageing women face much greater barriers to participating in leisure activities at precisely the time of life when such participation becomes more important to them.

In some cases, time use data suggest that retirement migration to coastal areas (see, e.g., Tomljenovic and Faulkner 2000) to adopt a perceived lifestyle based on leisure resources in the outdoors is not necessarily a reality. For instance, Wen *et al.* (2020) explored the concept of learned helplessness as a leisure constraint that creates a barrier to travel among an ageing population, with many studies culturally focused on Western contexts (e.g. Tretheway and Mak 2006). Given a proportion of the aged market live in poverty, access to leisure out-of-the home in commercial settings is far from equal. Ferrer *et al.*'s (2016) notion of social tourism – that is, 'tourism with an added moral value, of which the primary aim is to benefit either the host or visitor in the social exchange' (Minnaert *et al.* 2011: 414) – may help to overcome financial and personal barriers. Such trips are often funded by third-sector organisations (see Durko and Petrick 2013; McCabe *et al.* 2011).

Table 2.2 Selected elements of the American Time Use Survey, 2018: average hours per day (annual average all days of the week)

Age Group	Activity (Hours and Minutes)				
	Work	Leisure/Sport	Education	Household Activities	Personal Care/Sleep
15–19	1.22	5.79	3.06	0.84	10.34
20–24	3.52	5.02	1.70	1.12	9.96
25–34	5.01	4.22	0.28	1.14	9.54
35–44	5.08	4.08	0.09	1.13	9.28
45–54	4.93	4.54	0.04	1.17	9.30
55–64	3.79	5.49	0.05	1.18	9.40
65–74	1.05	7.17	0.03	1.28	9.65
75 and over	0.49	7.75	0	1.41	9.90

Source: Adapted from US Bureau of Labor Statistics, www.bls.gov/tus/

Table 2.3 Median gross earnings of employed in the UK, 2019

Age	Earnings (£)
18–21	16,830
22–29	25,057
30–39	31,812
40–49	34,633
50–59	32,038
60 and over	27,833

Source: Adapted from ONS (2019), www.ons.gov.uk/searchdata?q=Annual%20Survey%20of%20 Hours%20and%20Earnings

Having examined the leisure paradox, we now turn to developments in social theory, with a view to explaining the leisure–retirement relationship.

Leisure, retirement and ageing: social theory and leisure perspectives

Social theory, predominantly generated by sociologists, is a starting point to understanding the ageing–retirement–leisure nexus (Page and Connell 2010). Erikson's (1963) Life-Cycle Theory demonstrates that the older adult phase of life has many distinct features, and a great deal of theorisation has been devoted to how the ageing people spend their extra leisure time. As we saw in Figure 2.1, the range of leisure undertaken spans a continuum from in-home solitary activities to active out-of-home leisure and recreational pursuits to tourism (i.e. more than 24 hours spent away from home). For ageing people, this should be viewed in relation to the leisure paradox, measured in terms of rates of participation in specific activities (Hall and Page 1999; Page and Connell 2010). Lawton *et al.* (1995) observed that older adults tend to spend more time in and around the home, with very little sport and recreation undertaken (see also Kelly 1993). However, subsequent reviews argued that people's leisure pursuits remain relatively stable as they age (e.g. Patterson 2018). Moreover, in the'1980s, there was a growing realisation that retirement was also a time when individuals experienced new feelings of freedom to do what they wanted, when they wished, as well as providing an opportunity to take risks and try something they were unable to do while they were working' (Patterson 2018: 8), culminating in a positive view of ageing described by McGuire *et al.* (2013) as 'Ulyssean living'. As Patterson (2018: 8) explained, the Ulyssean concept dates back to the work of McLeish (1976), who described older generations' hunger for new experiences and adventure in later life. According to Adams *et al.* (2011), the notion that increased leisure enhances well-being in later life built on the lifestyle concept, while Tsartsara (2018) introduced the concept of geriatric tourism, which she defined as a positive sub-set of medical tourism.

Patterson (2018: 9), citing the work of Stebbins (1982, 1992, 1998, 2015) and the concept of 'serious leisure', noted that, 'for many older people, they achieved their greatest satisfaction and fulfilment from being amateurs, hobbyists, and volunteers, and this commitment helped them to keep busy, make new friends, and to enhance their older years'. This excellent synthesis of the leisure and ageing literature is particularly notable because it highlights the ageing consumer's broader engagement with the visitor economy, including tourism, in their leisure time. Yet, how should we theorise ageing within a leisure context to explain the patterns and activities of later life? Three theories emanating from sociology have formed the basis of many explanations of ageing, and each of them has its own strengths and weaknesses, according to Patterson (2018).

Activity Theory

Activity Theory, as articulated by Havighurst (1961), posits that remaining busy and active in retirement is the basis for achieving life satisfaction, complemented by raising a family. Specifically, Havighurst suggested,

> older people are the same as middle-aged people, with essentially the same psychological and social needs … The older person who ages optimally is the person who stays active and who manages to resist the shrinking of his [*sic*] social world … he [*sic*] maintains the activities of middle age as long as possible and then finds substitutes for those activities he [*sic*] … is forced to relinquish – substitutes for work.
>
> (Havighurst *et al.* 1963: 419)

Activity Theory was advanced by Lemon *et al.* (1972) and Longino and Kart (1982), who grounded their studies in a positive attitude towards ageing and later life. The basis of the theory is that remaining active in retirement, including social activity, helps to enhance one's self-concept. In retirement, an individual may substitute work-related roles with a range of new role identities. These new roles may include volunteering, recreation and involvement in the community, to achieve greater life satisfaction. Research into happiness in retirement (e.g. Menec 2003) also found that recreational activity improved well-being (Nimrod 2008). However, Patterson (2018) noted substantial critiques of this theory, such as Bowling and Gabriel's (2007) accusation that it was too simplistic to encompass the reality of later life and Burnett-Wolle and Godbey's (2007) contradictory conclusion that solitary leisure pursuits had a more beneficial effect. Similarly, Kleiber and Genoe (2012) highlighted the absence of a broader discussion of how mental health may be enhanced by more passive, informal and reflective leisure. Nevertheless, Alley *et al.* (2010) found that Activity Theory still featured prominently in numerous social gerontological studies.

Disengagement Theory

Disengagement Theory emerged in the 1960s from a study by Cumming and Henry (1961). Patterson (2018: 14) reported: 'this theory refers to the belief that as one ages there is a reduction or an abandonment of activities which was seen as a natural part of life, and this was in the best interests of the self and society'. Cumming and Henry (1961: 15) argued that this abandonment of activities means that older people become less engaged with life, other people and society, and their 'personal and social power declines' as a result. Moreover, this voluntary disengagement is compounded by the barriers faced in daily leisure. However, this negative view of ageing

> has now been largely discredited because several critics have suggested that this theory [Disengagement Theory] does not take into consideration the large number of older people who disengage voluntarily from those who disengage unwillingly due to a lack of support and available resources.
>
> (Patterson 2018: 15)

Even so, Alley *et al.* (2010) found that it was still used extensively in gerontological research.

Continuity Theory

In ageing research, Continuity Theory suggests that people will maintain most, if not all, of the activities, behaviours and social relationships that they undertook prior to retirement/later life. The main proponent of this theory was Atchley (1971, 1989, 1999), who built on the work of Maddox (1968). The underpinning concept is that people adapt to changing situations in the latter stages of the life course, with a focus on personality, ideas and beliefs. Other, external factors – such as relationships and social roles – provide a degree of support and stability in making these continuous adjustments. The argument in relation to leisure is that there is considerable continuity and consistency from childhood to later life.

Critics such as Strain *et al.* (2002) and Agahi *et al.* (2006) have highlighted inconsistencies and complexities in the application of Continuity Theory throughout the life course, not least because bodily functions and abilities inevitably change over time (see also Quadango 2007). Two of the main criticisms of the theory are its tendency to disregard the impact of chronic illness among older adults and its tendency to focus on male subjects. Nevertheless, Alley *et al.* (2010) reported that it was widely used in social psychology and featured prominently in studies such as Genoe and Singleton (2006). Later, Singleton (2017) acknowledged that ageing research will need to undergo further theoretical development if we are to fully understand the various challenges and cultural issues that global ageing will pose (see Iwasaki *et al.* 2007 for non-Western countries).

Irrespective of their various shortcomings, Alley *et al.*'s (2010) review indicated that the three theories discussed here are among the top ten most frequently mentioned approaches in the study of ageing (see Table 2.4 for a brief summary of each). However, the application of such theories in tourism studies remains challenging, since they tend to focus on marketing and segmentation as opposed to gerontology.

Table 2.4 Top ten most frequently mentioned theories in ageing research

Rank	Theory	Main Focus
1.	Life-course perspective	An interdisciplinary theory that focuses on the different factors that shape an individual's life from cradle to grave. This approach stresses the importance of cultural and social context in each individual's life.
2.	Lifespan approach	Emanating from Erikson (1963) and psychosocial theory, this approach focuses on eight stages of development over the course of the lifespan (i.e. infant, toddler, pre-schooler, school age, adolescent, young adult, middle-aged and older adult).
3.	Role Theory	Used extensively in sociology and social psychology, Role Theory suggests that everyday activity arises from socially defined roles (e.g. mother, manager, retiree). These roles are based upon a range of rights, expectations, obligations, norms and behaviours a person has to fulfil or address.
4.	Exchange Theory	Proposed by Dowd (1975), based on the idea that one has a perceived loss of status and power as one ages, and derived from the broader Social Exchange Theory, this approach argues that the 'social behaviour and interactions among individuals are a result of an exchange process. This perspective suggests that the relationship between individuals is generated by the pursuit of rewards and benefits and the avoidance of costs and punishment' (Wan and Antonucci 2016: 1). Social Exchange Theory was introduced by Homans (1958), who argued that individuals learn from previous interactions and consequently modify or maintain certain behaviours to achieve their main goals.
5.	Person–Environment Theory (and similar ecological theories of ageing)	This theory, sometimes termed 'environmental gerontology', has its origins in the deterministic view that old age is shaped by developments that are influenced and conditioned by the environment. As Wahl (2006) explained, it is contextual factors that determine everyday behaviour – a notion that can be traced back to the Chicago School of Urban Sociology in the 1920s. Therefore, the proximal influence of a positive or negative environment will shape one's life chances, health and life expectancy, giving rise to spatial inequalities.

(Continued)

Table 2.4 (Continued)

Rank	Theory	Main Focus
6.	Socioemotional Selectivity Theory	Initially developed by Carstensen (1992), this approach can best be described as a Lifespan Theory of Motivation. It suggests that we home in on social relationships and invest greater emotional resources in pursuing more meaningful goals and activities as our time horizon diminishes over the life course.
7.	Continuity Theory	Popularised by Atchley (1989: 183), 'Continuity Theory holds that, in making adaptive choices, middle-aged and older adults attempt to preserve and maintain existing internal and external structures; and they prefer to accomplish this objective by using strategies tied to their past experiences of themselves and their social world – change is linked to the person's perceived past, producing continuity in inner psychological characteristics as well as in social behaviour and in social circumstances. Continuity is thus a grand adaptive strategy that is promoted by both individual preference and social approval.'
8.	Activity Theory	Individuals remain busy post-retirement and in older age, with different activities bringing life satisfaction and happiness. Identifying with the new roles that novel activities provide boosts self-esteem.
9.	Disengagement Theory	This theory, which was formulated in the 1960s, views ageing as an inevitable process that is characterised by a decline in social interaction and engagement with society. Popularised by Cumming and Henry (1961), it was based on a series of principles associated with disengagement in old age.
10.	Feminist theories	Marshall (2006) traced the evolution of feminist studies of ageing to the early 1970s (e.g. Lewis and Butler 1972). Feminist scholars (e.g. Browne 1998; Gardner 1999; Ray 1996) analyse the lives of ageing women (including care-giving responsibilities and retirement) in a context of inequality, political economy, power, oppression and other critical issues that affect women in the latter stages of the life course.

Source: Adapted and developed from Alley *et al.* (2010)

The contribution of tourism studies to ageing research

Conceptualising tourism as a form of mobility

The concept of mobility is a good starting point for analysing ageing and tourism. Viry and Kaufmann (2015: 9) argue that geographers use the term 'mobility' to describe movement through space, which, of course, is a prerequisite for tourist travel, suggesting that mobility is organised around three axioms:

- *The range of possibilities*, determined by the availability of transport options, accessibility, the nature of the space and territories to be traversed, and the institutions and laws that govern human mobility (e.g. visa requirements, policies towards tourism), as reflected in concerns about pre-travel preparation (Ross 2005).

- *Mobility potential*, according to Viry and Kaufmann (2015: 9), depends upon an individual's life courses, family and career path. Their study highlighted that people have a certain mobility potential, which is also characterised by certain types of constraints (Viry and Kaufmann 2015: 9). This draws upon Crawford *et al.*'s (1991) and Godbey *et al.*'s (2010) studies of leisure constraints, which show that people are constrained and have distinct time–space prisms – a feature observed by Hägerstrand (1970). These prisms represent the ability of individuals and groups to move in time and space (i.e. geographic, economic and social space). Godbey *et al.* (2010) noted that coping strategies exist among the elderly, in that a reduction in participation means that they can substitute travel with other experiences (e.g. to reduce physical exertion), a feature also observed by Kazeminia *et al.* (2015) in the context of ongoing concerns around personal safety. So despite their having the leisure time, as Kazeminia *et al.* (2015) observed, ongoing concerns about personal safety (Lindqvist and Bjork 2000) did not lead to a growth in tourism activity. The constraints people face in their leisure time change in later life, as life events (e.g. becoming single after a partner's death) impact participation. In terms of tourism, visiting friends and relatives (VFR) becomes a key travel motive following bereavement, in order to offset loneliness (Von Soest *et al.* 2020). Parallel to this is cognitive decline in terms of memory, processing, reasoning and learning, which can make it difficult to assess multiple information sources and plan holidays. Older people tend to routinise purchasing behaviour, favouring pre-planned packages to reduce risk and derive benefits from VFR travel (see Carstensen *et al.* 2003; and Nielsen 2014 on negotiating constraints and motivational changes).
- *Movement*, which refers to geographic space. In tourism, this is largely a reciprocal action – departing from one's home area for more than 24 hours (e.g. Northcott and Petruik 2011) and returning in less than a year, but typically for one or two weeks of leisure (e.g. a holiday). McKercher (2018: 905) suggests that geographic space has an impact on travel in relation to distance, and 'distance has a profound, though often unappreciated impact on all aspects of tourism, extending well beyond the volume of tourist movements. It also reflects changes in the type of tourists who are most likely to visit a destination and their subsequent behaviour.' McKercher (2018) continues that the effect of distance in tourism diminishes as demand moves closer to supply. For example, 'land neighbours account for 57% of all [tourist] arrivals, while, collectively, destinations within 1000 km of a source market's border attracted 80% of all arrivals' (McKercher 2018: 905). Therefore, overcoming the barrier of distance remains a key element in tourist movement from origin to destination, along with other non-transport barriers that are associated with ageing and disability (see McKercher and Darcy 2018). We must also recognise that not everyone will avail themselves of the opportunities for tourism just because they exist.

As Dubois *et al.* (2015: 101–2) summarise, we need to think about mobility as

the set of personal characteristics that allows people to be mobile. It includes physical ability, aspirations to be mobile or sedentary, access to transport and telecommunications systems ... Mobility refers to (1) social conditions of access, that is the conditions that make it possible to use transport supply in a broad sense; (2) the skills required to use it; (3) mobility plans, that is, actual use of available transport.

The concept of mobility helps us to understand how ageing interacts with other characteristics and factors to shape individuals' or group's propensity to travel as tourists or for leisure travel. This new mobilities paradigm, which has also been termed the 'mobility turn' (Sheller and Urry 2006) in social science, places mobility at the heart of society as a phenomenon that shapes, impacts and redefines the societies in which we live. As a consequence of the increased speed of travel globally, time–space compression (Harvey 1990) has made much of the world accessible by modern forms of transport. Urry (2007) even argued that we should replace the focus of sociology as society with mobility in society. With the mobility debate in mind, we now turn to the tourism–ageing nexus.

Tourism studies and ageing

Alén *et al.* (2017: 1338) asserted that 'few studies analyse the relevance or motivations of senior tourists'. Conversely, Otoo and Kim (2020), informed by the literature on leisure, suggested that the majority of research on the elderly and tourism focused on travel motivation. They identified three separate categories of leisure studies: *user-oriented*, *activity-oriented* and *travel-oriented*. It is the latter category that we review here as a surrogate for tourism.

No theoretical approaches adequately explain the full diversity of travel experiences in relation to ageing. Many studies are anchored in the classic push-and-pull factors paradigm that characterised early critiques of tourism motivation (e.g. Crompton 1979; Dann 1981; Iso-Ahola 1980) and Pearce's (1993) travel career concept, which has been applied in different contexts and debated (e.g. Pearce and Singh 1999), culminating in a wide range of segmentation studies (Faranda and Schmidt 2000).

Otoo and Kim (2020), using the age of 50 as the starting point of ageing, reviewed 141 tourism studies on travel motivation published between 1980 and 2017, then narrowed down the selection to 36 key papers (see Table 2.5). This resulted in a list of 651 motivations, grouped into several domains (see Table 2.6). The top six motivations were destination appeal, socialisation, knowledge, novelty and escape and rest and comfort. The most important factors promoting ageing tourist travel were: VFR, cultural attractiveness, rest and relaxation, pride as a result of the visit/ability to tell others, exploration and curiosity. Tables 2.5 and 2.6 are based on Otoo and Kim (2020: 408), who propose a conceptual model (Figure 2.5), which built upon previous reviews of the field (e.g. Patuelli and Nijkamp 2016; Sie *et al.* 2016). Patuelli and Nijkamp (2016) focused on five key components (culture and nature; experience and adventure; relaxing; well-being and escape; self-esteem and ego-enhancement and socialisation) that were subsequently developed in Otoo and Kim's model.

Wellbeing
*Mental & physical wellbeing
*Sports/ physical invigoration
*Challenge & stimulation
*Health recuperation

Nostalgia
*Old friends/ family roots
*Memorable place attachment
*Relive memories
*Nostalgic feeling

Knowledge
*Intellectual enrichment
*Learning experience

Escape
*Feel safe/ Secure
*Escape stress/boredom
*Escape routine/ obligation
*Change diversion
*Escape environment

Ego
*Esteem
*Ego-centric
*Recognition/ respect
*Pride /Tell others
*Improve/ renew skills
*Feel privileged

Hedonism
*Fun of discovery
*Thrills or excitement
*Entertainment
*Pleasure

Drive motivations

Status motivations

Indulgence motivations

Actualization
*Self-fulfillment
*Self-reflection
*Self-reward /treat
*Peace of mind/ serenity
*Self-enjoyment/ happiness
*Spiritual/self-enrichment

Motivation

Rest & comfort
*Rest and relaxation
*Comfort
*Doing nothing/ Slow down

Novelty
*Experience of exotic
*Experience of nativeness
*Adventure/risk
*Experience of newness
*Exploration/curiosity

Supra-personal motivations

Socialization
*Family & friends
*Contact
*Interact/socialize
*Companion
*Connected/ community
*Share interests and values

Quality/ Specialization
*Shopping
*Tourism infrastructure
*Restaurants/accommodation
*Cuisine
*Personal concern
*Luxury

Destination appeal
*Cultural attractiveness
*Special events
*Heritage
*Natural/scenic environment
*Seek or escape weather

Travel opportunity
*Attractive transportation
*Value for money
*Health opportunity
*Time opportunity
*Recreation opportunity
*Information

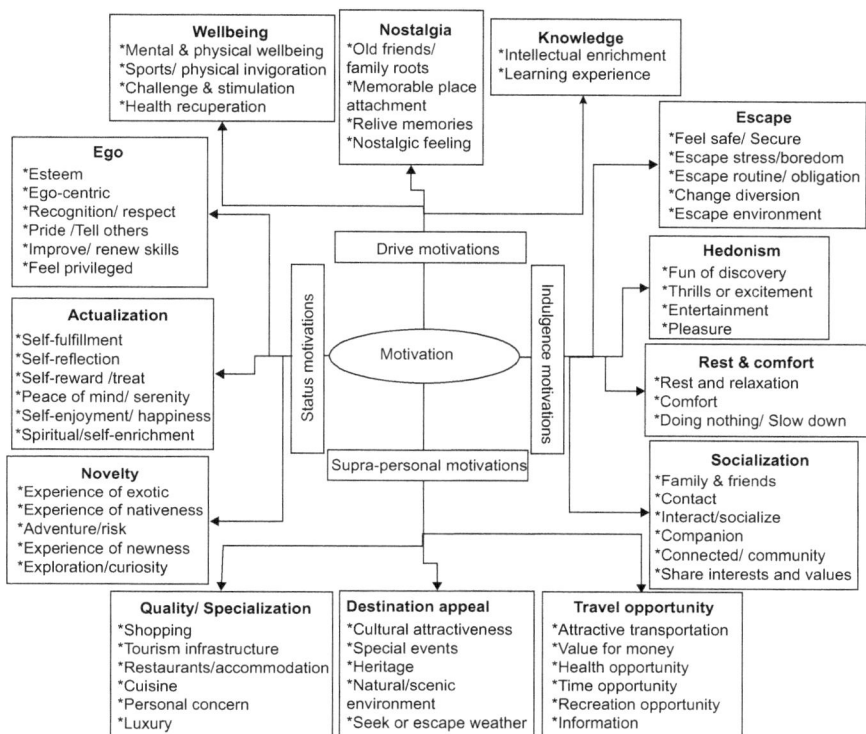

Figure 2.5 Conceptual framework for senior tourists' motivation

Source: Otto and Kim (2020)

Overall, these studies indicate that 'travel is regarded as an enjoyable leisure activity, and there are a multitude of reasons that help explain the travel behavior of older adults' (Sie *et al.* 2016: 100).

The travel experience can be analysed in three distinct parts (pre-travel, travel and post-travel), and Otoo and Kim's (2020) model illustrates that amongst ageing travellers pursuing educational travel.

- There are positive correlations between motivation and outcomes (i.e. health and leisure satisfaction);
- These benefits impact life satisfaction; and
- Positive experiences when travelling create positive memories and satisfaction from tourism.

Other studies in the ageing-tourism literature have explored Role Theory (e.g. Yiannakis and Gibson 1992) in the life course. González *et al.* (2009) examined the notion of the cognitive age of ageing travellers, challenging the use of calendar

Table 2.5 Publications on senior travel motivations

	Author(s)	Journal	Target	Study Setting	Minimum Age	Sample Size	Paradigm	Measurement Scale	Item Presentation
1	Carneiro et al. (2013)	Anatolia	Portuguese seniors	Portugal	60+	667	Quantitative	5-pointLikert scale	FA
2	Wang et al. (2017)	APJTR	Chinese urban senior outbound travellers	China	55+	360	Quantitative	5-pointLikert scale	FA
3	Musa and Sim (2010)	CIT	Older adults living in Malaysia	Malaysia	55+	1356	Quantitative	Descriptive	FA
4	Norman et al. (2001)	JHLM	Older individuals in selected states	USA	50+	374	Quantitative	4-pointLikert scale	FA
5	Janke et al. (2008)	JHTR	Senior attendees of education classes	Taiwan	65+	282	Quantitative	5-pointLikert scale	FA
6	Sellick (2004)	JTTM	Australian residents	Australia	50+	986	Quantitative	10-point Likert scale	FA
7	Hsu and Kang (2009)	JTTM	Chineseurban mature travellers	China	55+	800	Quantitative	5-pointLikert scale	FA
8	Prayag (2012)	JTTM	Senior travellers to Nice	France	50+	200	Quantitative	5-pointLikert scale	FA
9	Ryu et al. (2015)	JTTM	Older individualsin Japan	Japan	60+	25	Qualitative	In-depth interviews	Descriptive

(Continued)

Table 2.5 (Continued)

Author(s)	Journal	Target	Study Setting	Minimum Age	Sample Size	Paradigm	Measurement Scale	Item Presentation
10 Guinn (1980)	JTR	Elderly recreational vehicle tourists	USA	50+	1089	Qualitative	Descriptive	Descriptive
11 Baloglu and Shoemaker (2001)	JTR	Senior travellers from Pennsylvania	USA	55+	234	Quantitative	5-pointLikert scale	FA
12 Horneman et al. (2002)	JTR	Australian seniors	Australia	60+	724	Quantitative	4-pointLikert scale	Descriptive
13 Kim et al. (1996)	JVM	US senior citizens	USA	55+	914	Quantitative	5-pointLikert scale	FA
14 Muller and O'Cass (2001)	JVM	Australian 'young at heart'	Australia	55+	356	Descriptive	Descriptive	Descriptive
15 Boksberger and Laesser (2009)	JVM	Swiss senior travellers	Switzerland	55+	1101	Qualitative	Descriptive	Descriptive
16 Ward (2014)	JVM	Mature rich Irish individuals	Ireland	50+	266	Qualitative	5-pointLikert scale	FA
17 Sangpikul (2008a)	Tourism	US senior travellers	Thailand	55+	438	Quantitative	4-pointLikert scale	FA
18 Huang and Tsai (2003)	TM	Taiwanese senior travellers	Taiwan	55+	284	Descriptive	Descriptive	Descriptive
19 Jang and Wu (2006)	TM	Taiwanese senior travellers	Taiwan	60+	353	Quantitative	5-pointLikert scale	FA

(Continued)

Table 2.5 (Continued)

Author(s)	Journal	Target	Study Setting	Minimum Age	Sample Size	Paradigm	Measurement Scale	Item Presentation
20 Hsu et al. (2007)	TM	Chinese seniors	China	60+	27	Qualitative	In-depth interviews	Propositions
21 Lu et al. (2016)	TM	Chinese seniors	China	55+	360	Quantitative	5-pointLikert scale	FA
22 Borges Tiago et al. (2016)	TM	European seniors	Europe	65+	3458	Qualitative	Descriptive	Descriptive
23 Cleaver et al. (1999)	TRR	Australian retirees	Australia	56+	356	Descriptive	Descriptive	Descriptive
24 Chen and Gassner (2012)	JCTR	Chinese senior leisure travellers	China	55+	505	Quantitative	5-pointLikert scale	MANOVA
25 You and O'Leary (1999)	JTTM	Older UK travellers	UK	50+	405	Quantitative	4-pointLikert scale	Cluster
26 Lieux et al. (1994)	ATR	US residents	USA	55+	914	Quantitative	5-pointLikert scale	Descriptive
27 Kim et al. (2003)	TM	West Australian seniors	Australia	50+	200	Descriptive	Dichotomous	Descriptive
28 Shoemaker (1989)	JTR	Pennsylvania residents	USA	55+	407	Quantitative	5-pointLikert scale	FA
29 Shoemaker (2000)	JTR	Pennsylvania residents	USA	55+	234	Quantitative	5-pointLikert scale	FA

(Continued)

Table 2.5 (Continued)

Author(s)	Journal	Target	Study Setting	Minimum Age	Sample Size	Paradigm	Measurement Scale	Item Presentation
30 Alén et al. (2017)	CIT	Spanish senior travellers	Spain	55+	358	Quantitative	4-pointLikert scale	OVERALS analysis
31 Viallon (2012)	ATR	French and Chinese senior tourists	France/China	50+	564	Quantitative	5-pointLikert scale	FA
32 Eusébio et al. (2017)	JTCC	Portuguese citizens	Portugal	60+	848	Quantitative	5-pointLikert scale	FA
33 González et al. (2009)	IJCTHR	Older Consumers	Spain	55+	400	Quantitative	4-pointLikert scale	FA
34 Alén et al. (2014)	ATR	Spanish seniors	Spain	55+	358	Quantitative	4-pointLikert scale	Binomial mode
35 González et al. (2017)	JVM	Older tourists	Spain	55+	358	Quantitative	4-pointLikert scale	FA
36 Sangpikul (2008b)	IJTR	Japanese mature/older travellers	Thailand	55+	415	Quantitative	5-pointLikert scale	FA

Key: Factor Analysis (FA); *Tourism Management* (TM); *Journal of Travel Research* (JTR); *Journal of Travel & Tourism Marketing* (JTTM); *Journal of Vacation Marketing* (JVM); *Annals of Tourism Research* (ATR); *Current Issues in Tourism* (CIT); *Asia Pacific Journal of Tourism Research* (APJTR); *International Journal of Culture, Tourism and Hospitality Research* (IJCTHR); *International Journal of Tourism Research* (IJTR); *Journal of China Tourism Research* (JCTR); *Journal of Hospitality & Leisure Marketing* (JHLM); *Journal of Hospitality & Tourism Research* (JHTR); *Journal of Tourism and Cultural Change* (JTCC); *Tourism Recreation Research* (TRR).

Source: Otoo and Kim (2020: 402), with minor modifications

Table 2.6 Summary of senior tourist travel motives from key research studies

Domain	Factors Combining to Create the Domain
Socialisation	• Visiting friends and relatives • Contact • To interact/socialise • To share interests/values • Companionship • To feel connected/part of a community
Ego	• Pride in visit/to tell others • Esteem • Improve/develop travel skills • Recognition/respect • Feel privileged • Egocentric
Escape	• Escape routine/obligations • Escape stress/boredom • A change/diversion • Feel safe/secure • Escape environment
Destination Appeal	• Cultural attractiveness • Natural/scenic environment • Special events/festivals • Historic sites/museums • Seek good or escape bad weather
Knowledge	• Intellectual enrichment • Learning experience
Well-being	• Sports/physical invigoration • Challenge and stimulation • Health recuperation • Mental and physical well-being
Rest and Comfort	• Rest and relaxation • Doing nothing/slowing down • Comfort
Travel Opportunity	• Value for money • Recreation opportunity • Attractive transportation • Health opportunity • Information • Time opportunity
Novelty	• Exploration/curiosity • Experience of newness • Experience of risk/adventure • Experience of the exotic • Experience of indigenous lifestyles
Hedonism	• Seeking thrills or excitement • Entertainment • Looking for fun or discovery • Pleasure-seeking

(Continued)

Table 2.6 (Continued)

Domain	Factors Combining to Create the Domain
Quality	• Luxury • Cuisine • Restaurants/accommodation • Shopping/shopping facilities • Tourism infrastructure/facilities • Personal requirements
Actualisation	• Self-fulfilment • Spiritual/self-enrichment • Peace of mind/serenity • Self-reflection • Self-reward/treat • Self-enjoyment/happiness
Nostalgia	• Old friends/family roots • Memorable place attachment • Reliving memories • Nostalgic feelings

Source: Adapted from Otoo and Kim (2020: 406–7)

age as an explanation of motivation. Other studies have also aligned the social theoretical frameworks exploring life histories (e.g. Small 2003; Sedgley *et al.* 2006; Nimrod 2008) as part of a transformative research agenda proposed by Sedgley *et al.* (2011) to link tourism research on ageing more closely to critical gerontology. Embedding gerontology within research on tourism and ageing has not materialised. This reflects the tendency for tourism research over the previous 40 years to be 'confirmatory, reproductive … and formulaic', with qualitative inquiry underdeveloped (Sedgley *et al.* 2011: 422) and a reliance upon replication studies. The reason for these criticisms is evident in the ageing-tourism literature, with its notable absence of critical dialogue with gerontology. Instead, the social psychology of tourist behaviour literature on motivation seeks to address just one overarching question: what motivates the ageing person to travel for tourism? Few studies have diverged from this main focus to examine personalised accounts of ageing and thereby gain a more complete understanding of the precepts of 'emotion, agency, individuality … that promote the social inclusion, human dignity and human rights of older people' (Sedgley *et al.* 2011: 423), including humour (Mitas *et al.* 2012).

Research on ageing and tourism does not yet display the depth or breadth of insights that the geography of ageing demonstrates in terms of theoretical sophistication towards ageing, although one notable exception is den Hoed's (2020) analysis of age-friendly cycling and its impact on eudaemonic well-being (Nordbakke and Schwanen 2014). Den Hoed found a convergence between everyday mobility and slower, more age-friendly approaches to tourism – a feature also observed by Sheller and Urry (2006). This implies a need to design tourism and leisure experiences that have more meaning than just episodic escapes, which den Hoed (2020: 187) described as the pursuit of meaningful activity, personal growth and autonomy as

Mobility facilitates this pursuit, contributing to human flourishing by the places visited and by the act of movement itself ... A journey or an encounter while 'on the move' can thus improve well-being ... This relationship sharply contrasts the short-lived 'pleasure-oriented' nature of hedonic forms of well-being that is particularly important to older people ... this group is a core demographic in mobility research. As a population segment they are more wealthy and healthy than ever.

Banister and Bowling (2004) reached a similar conclusion. This fits within the agenda on active ageing (Hung and Lu 2016). The focus on segmentation was inherent in Faranda and Schmidt's (2000) study, which examined how the ageing process and bodily changes (both physical and psychological) occurred and how the market for tourism among older age groups was changing and becoming more heterogeneous.

Koskinen (2019) stressed the need to consider stages in the life cycle when designing tourism experiences with a view to supporting positive ageing (see also Bernini and Cracolici 2015). Yet, most ageing-tourism studies remain highly descriptive and translational (i.e. they apply general ideas and concepts in tourism to ageing contexts) and seem reluctant to engage fully with gerontology. Other studies include the pursuit of food tourism by ageing tourists (Balderas-Cejudo *et al.* 2019) and marketing to ageing tourists (e.g. Patterson and Pegg 2009a), while Chen and Shoemaker (2014) explored age and cohort effects in ageing research, drawing upon Generation Theory, Continuity Theory and Life-Cycle Theory. Generation Theory focuses on generations and their characteristics, and it has informed the development of research on generational marketing. In terms of the visitor economy, this generational marketing focus led to a focus on baby boomers (e.g. Borges Tiago *et al.* 2016 and Cejudo 2018), who were perceived as avid travellers. These studies built upon those by Glover and Prideaux (2009) and Bieger and Laesser (2002). Le Serre and Chevalier's (2012) research into marketing travel services to the aged also drew heavily upon segmentation and motivation studies, while Lehto *et al.* (2008) compared baby boomers with the silent generation. Wang and Feng (2014) explored the demands which all of these groups place on tourism producers.

Other popular subjects include accessible tourism, since around 8% of ageing tourists may have access needs, given that disability tends to increase with age (Darcy and Dickson 2009; see also Chu and Chu 2013; Wang and Feng 2014). The arguments for accessible destinations outlined by Darcy *et al.* (2010) have not seen any take-up in academia or industry to any significant degree. This is because the research agenda has now moved towards new issues such as age-friendly cities (see Chapter 7). Instead, industry efforts have focused on the wider implementation of disability discrimination legislation and equality acts that encompass the needs of an ageing population, particularly in nations with pronounced ageing issues, such as Japan (Mak *et al.* 2005; Chen *et al.* 2018; Funck 2008). Yet, as den Hoed (2020) indicates, principles of universal design are key to creating accessible destinations (see Chapter 3). De Guzman *et al.* (2019: 658) drew upon positive psychology in their study of mindfulness and ageing (see Langer (2000) on mindfulness), suggesting, 'Mindfulness refers to the tourists' consciousness of their surroundings, time, activities, belonging and company driving a guided tour. Tourists' involvement concerns the extent to which tourists participate actively in the activities included in the guided tours' and how they are engaged in the present (Langer 2000). Studies of these areas have also looked at issues around happiness and fulfilment (Gershon and Gershon 2000). The numerous country-specific studies on ageing and travel (e.g. Grundey and Vilutyte 2012; Möller *et al.* 2007; Pochun 1999; Romsa and Blenman 1989) tend to focus on barriers to travel, such as securing personal insurance (e.g. Henderson 2007), and/or particular modes of transport, such as coach trips (e.g. Hsu *et al.* 2007).

Other studies have investigated health tourism, which Tretheway and Mak (2006) described as a niche sector for what is termed perhaps somewhat controversially 'geriatric tourism' in developing economies. Tsartsara (2018) studied the use of medical tourism facilities, especially among the over-70s. The tourists who visit such facilities may be characterised as having mild cognitive impairment as well as physical and psychological frailties. Other studies have examined differences in the geography of ageing holidaymaker behaviour (Losada *et al.* 2019), the role of shopping (Littrell *et al.* 2004) and the neglected area of lesbian ageing (Gabbay and Wahler 2002). In motivation-focused studies, there has been a growing recognition of a correlation between driving in old age and increased risk of accidents and injuries as cognitive functions decline (Gaugler 2014). This correlation has been confirmed in studies of self-drive ageing tourists (e.g. Prideaux *et al.* 2001). Finally, several studies have looked at policy developments in certain countries (e.g. China) to facilitate tourism for their ageing populations. These highlight the importance of understanding how organisations classify and characterise the ageing consumer.

Tourism as a consumer activity: Life-Cycle Theory, generational marketing and ageing

As Huber (2019: 372) indicated, 'as people grow older, they are increasingly confronted with physical and psychological developments and changes in their social environment … The variety of life course transitions in older adults' lives suggests that seniors develop tourism patterns that mirror these living circumstances.' In other words, life-course changes impact people's motivations, ability and engagement with leisure and tourism. The myriad developments that occur during the natural ageing process led Schewe (1988) to list many of the physiological changes that people undergo in later life, some of which are presented in Table 2.7. Schewe (1988) argued that most products were designed for younger consumers, and by bringing gerontological research to a consumer research audience, highlighted a need for greater attention to the ageing consumer market, drawing upon generational marketing techniques.

Table 2.7 Selected changes resulting from human ageing

Change to Physical Characteristic (Increase)	Change to Physical Characteristic (Decrease)
• Ear length and breadth • Dental alterations • Greying hair • Pelvic breadth	• Sitting height • Arm span • Muscle mass • Skin elasticity • Range of motion • Bone density • Body weight • Vision • Hearing capacity

Source: Adapted from Schewe (1988)

Since Schewe's (1988) call to action, a range of marketing studies and a lesser number of tourism studies adopted the model of the life course as consisting of distinct eras in one's life. Nimrod (2008) and Huber (2019) examined tourism practices directly impacted by health, income and availability of travel partners in later life. Similarly, the dependence upon caregivers, after ageing leads to failing health, may curb tourism by adding further barriers to travel, as is the case with people with dementia. Supporters of the life-course proposition assume that specific life events, such as retirement or illness, make travel in older age much more complicated (Huber 2019), so qualitative approaches are needed to appreciate the nuances in individuals' life histories and paths. Huber (2019: 384) concluded that the portfolio of tourism behaviour is impacted by life events, resulting in complex interactions between heterogeneous and homogeneous behaviours. This challenges existing thinking around unicausal relationships between ageing and travel behaviour as the various motivations discussed earlier create a heterogeneous tourism market.

A range of marketing texts have adopted the life-course concept to investigate the ageing consumer. For example, Coughlin (2017) indicated that the ageing consumer was worth US$8 trillion in consumer spending globally, of which spending in the visitor economy was a key component. Coughlin pointed to women's increasing role in leading spending in older age groups, building on Schewe's arguments that new physiological, cognitive and psychological traits (e.g. loneliness in single-person households) as well as familial trends impact spending among ageing consumers. In the UK, longitudinal studies such as the Family Expenditure Survey have recognised that the consumer experiences of today's ageing population will differ substantially from those of their predecessors (Jones *et al.* 2008). Stroud (2005) went so far as to propose age-neutral marketing in order to improve communication with this market. In Asia, Hedrick-Wong (2007) examined the rise of consumer markets and their clustering in megacities, while a more age-friendly thesis was proposed by Stroud and Walker (2013). However, it would appear that many conventional consumer behaviour-focused studies (e.g.

Le Serre and Chevalier 2012) rely upon the conventional marketing techniques of segmentation and motivation. Even so, in some regions (e.g. Africa), ageing remains a problem area for which policymakers and business are unprepared (e.g. Makoni 2008), with the voices of the ageing largely silent in most narratives (Makoni and Stroeken 2017).

While many studies on the consumer behaviour traits of ageing consumers adopt a generational marketing orientation, more gerontologically informed analysis is evident in the concept of gerontographics advocated by Moschis (1996) and Moschis and Ünal (2008), based on the principles of psychographics and lifestyle segmentation. Even so, few studies adopt a qualitative focus, as illustrated by Nimrod and Adoni (2006), Nimrod (2008) and Nimrod and Rotem (2011). Building upon Nimrod and Kleiber's (2007) innovation theory of successful ageing, Nimrod and Rotem (2011) observed the importance of novelty in the behaviour of ageing tourists who are willing to try new travel services and holiday products. Nimrod and Kleiber (2007) complemented existing social theories of ageing, arguing that leisure behaviour is often associated with a specific behaviour (i.e. involvement in a new leisure activity) that may have positive psychological outcomes (self-preservation or self-reinvention, enhanced sense of meaning in life and well-being) (Nimrod and Rotem 2011: 385). These studies highlight that new marketing approaches in consumer behaviour have converged to examine the experiential aspects of ageing tourists as consumers. This extends a long history of research that emanates from leisure studies (see, e.g. Adams *et al.* 2011; Scherger *et al.* 2010; Osgood 1982; Leitner and Leitner 1995; Siren and Haustein 2015). Research on ageing has used different leisure methodologies such as oral history to understand life histories (e.g. Thompson 1992), time budget analysis (e.g. Little 2008) and similar methods of analysing time use (e.g. Ekerdt and Koss 2016), longitudinal studies (e.g. Mollenkopf *et al.* 2011) and a focus on individual lives (Roadburg 1985) and the effects of life events (e.g. surrendering one's driving licence; Siren and Haustein 2015) to build on seminal studies such as Lemon *et al.* (1972). Lemon argued that cultural norms in Japan, where there is a strong group orientation in employment setting, makes it difficult for individuals to prepare for retirement and the individualism that arises post-retirement (McCallum 1988).

Evidence from the leisure literature continues to highlight the importance of contextualising tourism in terms of leisure experiences either post-retirement or in later life and for those who remain in some form of employment (Harahousou 2006). There has been debate within the literature on the extent to which leisure and tourism behaviour displays a degree of continuity or change through the life course (e.g. Strain *et al.* 2002; Agahi *et al.* 2006). Godbey *et al.*'s (2010) review of the earlier models of leisure constraints research (Crawford and Godbey 1987; Crawford *et al.* 1991) suggested that the main reasons why the degree of continuity might change in later life relate to the development of structural, intrapersonal and interpersonal constraints as we discussed earlier. A good example of this is the study by Innes *et al.* (2015) in relation to dementia and the visitor economy which was framed in terms of these constraints and barriers to leisure. Later life events, such as becoming single on the death of a partner, impact participation as well as the propensity to visit family and relatives, although 'it would be naïve, however, to expect that all individuals – in all social, cultural, and historical contexts – would experience the same set of constraints and perceive each of them to have the same importance or strength' (Godbey *et al.* 2010: 119). Therefore, given the ageing tourist is a consumer, attention now turns to the affordability of holidays and financial considerations.

The financial climate and affordability of leisure and holidays: the pensions crisis and retirement

From a demographic perspective, one might say that the world's ageing population is in the midst of a period of hypergrowth (Koskella and He 2008). For example, it took over a century for France's over-65s to increase from 7% to 14% of the total population. In contrast, China is expected to achieve a similar transformation of its age profile in 25 years, and South Korea in just 18. To put this into perspective, this would amount to a numerical increase from 110 million over-65s in China in 2011 to 330 million by 2050. Koskella and He (2008) extrapolated this to an additional 810,000 over-65s around the world every month for the foreseeable future. This reflects the direction of travel in demographic transition theory as population pyramids become more rectangular, the distribution of population by age group becomes more evenly spread and the increase in life expectancy means there are ever more people in older age groups.

With these demographic changes in mind, the ability of the ageing consumer to engage with the visitor economy will be significantly affected by the availability of financial resources, alongside other factors that facilitate or inhibit participation in the visitor economy. In view of the relationship between retirement and ageing, the reduction of income is a common thread in many debates about post-retirement and later life. Ensuring sustainable forms of post-retirement income to replace income derived from working means pensions remain an important source of replacement income in retirement. In the UK, 2 in 3 people in retirement derive income from private pensions, and among 50% of those receiving pensions, their incomes are aligned to the income distribution of the general population, based on government data in the Pensioners Income Series (www.gov.uk/government/collections/pensioners-incomes-series-statistics – 3). However, the other 50% do not enjoy the same benefits of income security from private pensions and so are dependent upon government support. An interesting analysis by Thomson (1984) noted the relative decline in the value of pensions since the Victorian and Edwardian period, a finding which challenges much of the conventional wisdom on the beneficial effects of a welfare state in the UK. The main sources of income in retirement are state benefits, state pensions, private pensions and investment income. At a global scale, a number of common trends are impacting pensions. According to the OECD (2019), in OECD countries the ratio of people of working age to those over 65 stood at 10:2 in 1980 and 10:3 in 2020, but it is projected to shift to 10:6 by 2060. This reflects a more general decrease in the proportion of working population alongside the following trends:

- Significant increases in people of pension/retirement age;
- The reform of many state pension schemes, typically in the form of increased contributions to achieve greater financial sustainability;
- A shift towards expanding private savings and pensions alongside incentivising people of retirement age to remain in work; and
- Greater alignment of public sector pension schemes with their private sector counterparts.

Yeh *et al.* (2018) examined the implementation of these policy shifts in Asia to illustrate the complexity of pension models around the world. As the global review by Pensions Watch (2020) illustrated, 114 countries currently have pension provision, although this is notably poor in some regions (e.g. Africa and the Middle East). For example, only eight countries in Africa have full pension provision. Mercer (2019) highlighted the pressure on pensions given the low-growth/ low-interest-rate environment that is currently impacting financial returns, as well as the rise of defined contribution schemes, which have started to replace defined benefit or final salary schemes because the latter are largely unsustainable on account of the general rise in life expectancy. In addition, the OECD (2019) raised concerns over the growth of the gig economy (i.e. a labour market with short-term, zero-hours and freelance contracts) and intermittent earnings (due to non-standard income), which now account for a third of jobs in OECD countries.

The sustainability of state pension schemes is examined in Mercer's (2019) Global Pension Index, which covers 37 countries and 62% of the world's population. Just two countries were rated 'robust' (the Netherlands and Denmark); a small group had *sound pension structures* (including Australia, Canada and New Zealand); a much larger group had *some good features* as well as a range of risks and shortcomings (including the UK, Hong Kong, the USA, France and Poland); and a final group had *some desirable features* but faced significant long-term sustainability issues (including Japan, South Korea and China).

With specific reference to tourism, reliable incomes from pensions and private means have significantly benefited many members of the baby-boomer generation – retirees as well as people who have remained in work – especially as the cost of overseas travel has dropped substantially since the 1960s (see Foley and Rhodes 2019). In addition, this generation has benefited from further cost reductions due to the rise of low-cost airlines, the disruptive effects of technology (e.g. Airbnb) and governments' reluctance to tax aviation fuel. Yet, these benefits have not been uniformly distributed. Foster (2018) pointed out that there is an urgent need to shift the debate on the relationship between ageing and pensions away from its current, almost exclusive focus on prolonging employment and postponing retirement so that proper consideration can be given to active ageing. There is also a recognition that gross inequalities in retirement income reflect wider disparities in societies that are split between the 'haves' and the 'have-nots'.

Poverty and old age: the 'haves' and the 'have-nots'

The visitor economy is largely premised upon the concept of consumer spending – that is, paying for leisure or tourism-related activities with disposable income (although daily leisure may be non-commercialised). Therefore, as the previous section illustrated, the availability of disposable income is key to understanding the likely demand for products and services in the visitor economy. According to the UN Department of Economic and Social Affairs (2016), in regions where poverty is endemic, those who survive a lifetime of poverty often face an old age of even greater poverty. This is reminiscent of the findings of Booth and Rowntree in Victorian England (see Preface). While data on poverty in old age is limited, those in the less developed world suffer more than those in the developed world. Nevertheless, the UN Department of Economic and Social Affairs (2016) found that 14.7% of over-75s in OECD countries were living in a state of poverty, compared with only 11.2% of 66–75-year-olds. Hence, it seems that poverty increases with age, even in the world's richest countries.

Two crucial components of poverty in old age are household indebtedness and unreliable retirement income. These need to be considered in the context of many governments' promotion of financial self-reliance in retirement. As Silcock (2015) indicated, future retirees up to 2040 will be ever more reliant on defined contribution pensions, which almost invariably pay less than the former defined benefit schemes, so they are likely to have significantly lower incomes in old age than baby boomers are currently enjoying. In simple terms, many governments are concerned that future pensioners will soon exhaust their savings due to these less generous schemes, longer life expectancy and low levels of savings. In addition, Lloyd and Lord (2015) suggest that a natural decline in cognitive ability can propel pensioners into financial hardship as they start to lose the economic knowledge and experience they accumulated earlier in life.

Addressing this issue is extremely complicated, not least because there is no universally accepted, precise definition of 'poverty', although it is generally agreed that it constitutes a lack of sufficient income to support basic needs (e.g. food, clothing, shelter, heating and personal care). Hence, those who live in poverty cannot participate in activities that those with greater affluence consider to be routine aspects of day-to-day life. Meanwhile, the still widely used terms 'extreme poverty' and 'destitution' have strong associations with Victorian attitudes to pauperism and the state workhouse system.

Many classic studies (e.g. Townsend 1979; Mack and Lansley 1985) have reviewed the various definitions of 'poverty' and the measures that have been used to gauge it, including:

- *Absolute poverty* – that is, extreme poverty in which hunger, potential death and human suffering are acute.
- *Relative poverty* – an income of less than 60% of median income, which is likely to lead to hardship. (Some organisations, such as the Joseph Rowntree Trust, have preferred slightly different measures, such as 50% of mean household income.)

- *'How poor is too poor?'* – the question Mack and Lansley (1985) attempted to answer in order to determine a level of income and resources that would lift an individual or household out of deprivation.
- Townsend's (1979) concept of *relative deprivation*, which considered how people lived and the resources that were necessary to achieve a minimum standard of living on the basis of 60 indicators. Indeed, Mack and Lansley (1985) criticised this approach on account of Townsend's seemingly arbitrary choice of indicators.

In addition, governments have utilised the concept of *social exclusion*, where the emphasis is on how individuals and households are denied access to resources, limiting their ability to participate in society. This concept is imbued with notions of equity, social cohesion, the effect on one's quality of life and how poverty and deprivation affect social, economic and cultural participation.

Whichever approach or methodology is used, there is agreement that the creation of wealth, its distribution, and the political structures that may alleviate or compound inequalities in that distribution, are fundamental features of capitalist society. In business, minimising labour costs can enhance profits and make a company more competitive in a free-market economy (Stiglitz 2020), but the impact of this on pension provision has emerged as an important issue in tandem with ageing. The distinction between the 'haves' (i.e. the affluent and those who have benefited from the capitalist system) and the 'have-nots' (i.e. those who have not benefited) becomes more evident in old age, since the 'haves' are able to accumulate wealth and assets (e.g. a property or properties) and make provision for their retirement during the course of their working lives. In contrast, the very poor 'have-nots' will typically have no such opportunity to make these provisions.

This rudimentary way of examining income inequality and wealth distribution by considering the 'haves' and 'have-nots' is a useful way to illustrate a significant gulf in society. As Rosenteil (2007) indicated, between 1987 and 2007 in the USA, the proportion of 'have-nots' doubled from 17% to 34% of the population, based on individuals' perceptions of their lot. Similarly, between 1980 and 2004, the top 19% of households' share of national income rose from 8% to 16% in the USA.

Poverty, when it is measured, remains an enduring theme for many capitalist societies, especially among the older population, just as it has been for more than a century. A report by the Joseph Rowntree Foundation (2017) tracked this persistence in poverty and older people in the UK between 1994 and 2016. It found that old age poverty rose from 13% in 2012/13 to 16% in 2017 – a 300,000 increase in absolute numbers – despite sustained investment by the state to alleviate pensioner poverty through pensions and income support. Similarly, Age UK (2019) stated that 16% of all pensioners were living in poverty in the UK, although this figure rose to 35% for private tenants and 29% for those in social rented sector, compared with 13% of those who owned their home. In terms of ethnicity, 31% of Asian/Asian-British pensioners and 32% of Black/Black-British pensioners were living in poverty, compared to 15% of their White-British counterparts. Finally, disability may exacerbate financial insecurity and exclusion from many activities in old age.

If wealth and poverty are two ends of the spectrum that impacts everyone's participation in the visitor economy, how do pensioners spend their income on tourism? Thus far, Pak (2019) has been the only researcher to address this issue. In a review of Zhao *et al.*'s (2016) work, Pak found that pension income can create a high marginal propensity to consume. Focusing on South Korea, he suggested that pensioners used 'additional pensions to achieve the desired level of tourism consumption' after basic pension reforms targeted additional income to 70% of the country's pensioners (Pak 2019: 12). Travel frequency increased from 23% to 59.5% and spending from 52% to 96%, dependent upon pensioners' socio-economic profile. Pak established that 45% of South Korean pensioners had been living below the poverty line in 2014, but about half of these people became tourists following the country's pension reforms, with consequent well-being benefits.

Morgan *et al.* (2015) acknowledged that retirement accentuates wealth disparities, since these intensify with age, while Casey (2012) forecasted that turmoil in the financial markets and reductions in pension payouts will make holidays less affordable for some retirees. Reflecting on a town with a low-income profile worked in prior to retirement, one of our respondents (R1) argued:

> those people who are still part-time working have less leisure so those people in their early 70s now who are impacted by later working don't have as many opportunities … But the people who are properly and fully retired often have the leisure but not the income necessarily to support doing the things they want, so that's the thing. In terms of addressing the differences, the main thing is physical disability or cognitive impairment. There are several generations in there – from people that are sort of active and retired through to people that have been retired ten years and [are] maybe a little bit out of touch, or maybe they still work part-time because, again, ageing is changing … The increase in retirement age [had] a dramatic [impact on] the way that people's lives changed. So something like one in ten people in their early 70s still works part-time, and that's extraordinary because that keeps them more connected to the outside world and that might well have [an] impact down the line. In a pre-Covid world, I mean – post-Covid it's entirely different – but in a pre-Covid world that would have kept them more connected to being in touch with things, whereas for people who have retired and they're out of the workforce and they're perhaps not volunteering, they're perhaps more at home, the world is a scarier place because they're getting a lot of their messages through the television and radio and newspaper, you know – 'Don't trust young people', 'Foreigners are taking over the world' – and, you know, everything is doom and gloom and despondency and it's very frightening and they can become quite reactionary in their views. They become quite afraid.

Thus, working longer appears to have an impact on the ageing population's time for leisure as well as their engagement with society and staying connected, while the Covid-19 pandemic has had a detrimental impact on both perceptions of the world and how views are shaped.

Morgan *et al.* (2015) pointed to the challenges that older people face in terms of loneliness and isolation, a subject to which we will return in later chapters, and indicated that tourism and leisure participation may help to address these issues. They interviewed participants in a social tourism programme – the National Benevolent Fund for the Aged Break Away Scheme – who reported numerous well-being benefits from taking a holiday, such as: the respite it provided from mundane day-to-day life; reductions in feelings of marginalisation; increased self-esteem; and help in coming to terms with the loss of a partner. This study is important as it highlights the benefits of tourism for the psychological well-being of ageing consumers – a key element in the discourse on ageing well to lead lives that are socially rewarding, have meaning and are not blighted by loneliness (Age UK 2011) (see Chapter 5 for a discussion of isolation and loneliness). As Morgan *et al.* (2015: 3) pointed out, 'older people frequently have to deal with often interconnecting life transitions such as bereavement, physiological change, increased ill-health and reduced socio-economic circumstances'. It is against this backdrop that we now turn to the subject of active ageing.

Active ageing and participation

Walker (2015: 14) argued that 'active ageing is now established as the leading global policy strategy in response to population ageing'. This impacts the ageing–visitor economy nexus since the ageing population's use of leisure time is typically structured around at-home and out-of-home leisure, where many daily and bi-daily experiences interact with the visitor economy. As Walker (2015: 14) astutely stated, 'active ageing often serves merely as a convenient label for a wide range of contrasting policy discourses and initiatives concerning ageing and demographic change'. Walker traced the evolution of the active ageing concept to discourses by international government organisations like the World Health Organization in the 1970s, but the idea really gathered momentum in the 1980s as developed countries started to recognise the economic costs of ageing (Walker 1980; Townsend 1981). As Walker (2015) pointed out, the narrative of social movements in civil society (which we will examine in Chapter 3) illustrated the fact that increasing numbers of ageing people and longer life expectancy were starting to challenge the main tenet of the welfare state – to provide a safety net for all through pensions and benefits.

Active ageing emerged in reaction to social theories of ageing, focusing initially on Rowe and Kahn's (1987) model for successful ageing, which expanded upon Pfeiffer's (1974) collection of papers. As Walker (2015: 18) explained, Rowe and Kahn's model hinged upon three key principles: (a) a low probability of disease in old age and disease-related disability; (b) the retention of high levels of cognitive and physical capacity; and (c) active engagement in later life. Despite critiques that pointed to difficulties with this concept, its positive outlook stimulated policy debate, with the WHO (1994) promoting the concept of healthy ageing, which was also rolled out by bodies such as the European Union. The principles in this iteration of active ageing identified a clear link between activity and health, promoting participation of an ageing demographic to achieve inclusivity across the population. Walker (2015: 19–20) stated, 'the thinking behind this new approach ... is expressed perfectly in the WHO dictum "years have been added to life now we must add life to years"', recommending the promotion of more active lifestyles to maintain and enhance the physical and mental health of an ageing population as embodied in the concept of quality of life (QoL). As Walker summarised:

> active ageing in this conceptualisation concerns the optimisation of activities related to a wide range of endeavours: employment, politics, education, the arts, religion, social clubs ... increasing the unpaid and paid contributions older people make in society challenges views of older age which emphasise passivity and dependency, by alternatively emphasising autonomy and participation (Walker 2015: 2020).

The concept of active ageing promotes adoption of a holistic approach to the lives of the ageing population, understanding the factors and barriers that can be addressed to enhance well-being. Walker (2015) identified a number of these barriers:

- A *political barrier* existed due to confusion over the term active ageing, even though it was often used for political ends. First, as is evident in the grey and academic literature, multiple terms and concepts are often substituted for 'active ageing' (e.g. 'positive ageing', 'healthy ageing', 'successful ageing'), all of which simply mean ageing well. The implementation of any strategy to promote active ageing requires the engagement of multiple stakeholders. Second, ideological misuse of the term has arisen to encourage the ageing population to remain healthy in order that they might work longer, as the age of retirement has been increased in many countries.
- *Cultural barriers*: the stereotypical promotion of a fit, athletic and highly engaged pensioner may be misleading to the very people active ageing strategies are targeting.
- *Bureaucracy*: the holistic ideology implicit in developing active age strategies can hit the buffers due to the involvement of multiple government departments, requiring resource-sharing, non-silo thinking and addressing issues of advocacy.
- *Societal*: ageism and age discrimination (direct and indirect) stigmatise older people, creating stereotypes and imagery that negate active ageing strategies and portray them as of limited benefit, reinforcing marginalisation.
- *Unequal ageing*: categorisation as a 'have' or a 'have-not' is established early in an individual's life course, whereby inequalities develop and are reinforced and perpetuated across the lifespan, although they are most visible in old age.

As Walker (2015) pointed out, 'one size fits all' active ageing strategies tend to overlook these inequalities.

To address these barriers, Walker (2015) advocated eight principles for developing a comprehensive and consistent active ageing strategy:

1. Activities included under the banner of active ageing should have meaning to the individual/family concerned, such as volunteering.
2. The concept should be preventative, involving all age groups and focused on maintaining functional capacity. When a disability threshold is crossed, active ageing must ensure capacity and capability are maintained for as long as possible to enhance quality of life.
3. Active ageing should be inclusive and involve all groups, ranging from the 'young-old' to the 'old-old', including those who are frail and dependent.
4. Maintenance of intergenerational solidarity to involve ageing relatives in activities.
5. Rights to all forms of social protection, training and education must be respected throughout the life course.
6. Active ageing should be empowering and participative.
7. Cultural and national diversity must be embedded in any strategy to reduce the problem of unequal ageing.
8. Active ageing must be flexible.

In addition, all of these principles must be based on the notion of a partnership between the citizen and society.

Various studies have reviewed the implementation of active ageing in different countries (e.g. Walker and Aspalter 2015; Zaidi *et al.* 2018; Boulton-Lewis and Tam 2012; Timonen 2016; Riva *et al.* 2014; Formosa 2019; Hofäker *et al.* 2016; Baskaran 2020; Walker 2018) as it has become a very convincing ideology that now permeates many policy domains, interconnecting with civil society, as we will illustrate in Chapter 3.

The multigenerational family and tourism

One important demographic change that longer life expectancy is creating is the multigenerational family (MGF). According to Generations United (2011: 3), the United States Census Bureau defines an MGF as 'one which contains three or more parent–child generations'. In 2010, 3.8% of US households were in this category and the figure was increasing. There are, however, alternative definitions, such as: an adult living with a parent; a couple with their adult child, who has returned to live in their household with them after a period living elsewhere; a couple with a parent; or a grandchild and grandparent living in the same household. In the USA, when such broad definitions are included in calculations, the proportion of MGF households stands at more than 10%, and it is even more prevalent among the country's Hispanic and Asian populations. According to Generations United

(2011), the growth in MGF households is primarily motivated by compassion, but socio-economic factors play a role too, including:

- Couples marrying later;
- Emigration from cultures where MGF living is more prevalent (in the USA, 19% of immigrant households may be categorised as MGF, compared to just 14% of native-born households);
- The financial security of many members of the baby-boomer generation, who have been able to support their parents; and
- A rise in the number of ageing people with disabilities and/or chronic health conditions who receive care from family members.

In the UK, the Office for National Statistics (ONS) has estimated that multi-generational households (MH; an alternative term for MGF) increased from 325,000 in 2001 to 419,000 in 2013, but the actual figure may be as high as 1.8 million, depending upon the criteria used. Studies in the USA suggest MHs account for about 20% of the population (Trapper 2019).

From a tourism and leisure perspective, MHs pose theoretical issues around the nature of the experience and the relationships it engenders. In addition, there are practical issues relating to new accommodation configurations with activities for all generations, including baby boomers as grandparents (Ruspini and Del Greco 2017). VisitEngland described the characteristics of this market in its *One Minute to Midnight* report (VisitEngland 2019), typically in parent–child configurations (which are most prevalent among French travellers) that included at least one ageing parent or other relative. Various other market intelligence studies have highlighted the volume of MH travel, with

- 20 million Americans taking an MH trip in 2011;
- 33% to 40% of all leisure travel in the USA based on MH travel;
- More families who live far apart using MGF holidays to reconnect with one another; and
- The proportion of MH holidays increasing annually.

However, multigenerational holidays are not without their problems, as one of our respondents (R1) noted:

> When people do have families, normally it will be the younger adults in the family who will book the villa, arrange the flights, arrange the transfers, whatever … so disabled-friendly [aspects] and so on … people don't think about. So it's not uncommon to get older people to say, 'Oh, my daughter booked a villa in France and I was stranded there the whole time. I couldn't leave … It was beautiful but once we actually got there, there were too many hills, there was this, there was that, I couldn't get out and about.'

Heimtun's (2019) analysis of taking holidays with ageing parents introduced the concept of 'filial duties' – that is, 'adult children's obligation towards their parents' – including a wide range of activities from care-giving as part of an MH or in their own home through to visiting them and taking holidays with them to strengthen bonds. Heimtun (2019: 136) used the word 'intergenerational' when one adult and child were involved, arguing that

> filial duty thus changes across the life course. Major transitions occur when parents' health declines and when one parent dies. Death and changes in an ageing parent's health, particularly with a resultant role reversal, put extra pressure on some of the participants, who slowly felt 'locked into' filial duty during holidays.

Ageing and care settings: housing needs, retirement villages, sheltered accommodation and care homes

An early study of leisure activities in retirement by Sherman (1974) found greater engagement in leisure activities in retirement homes than when people continued to live in their own houses. With regard to the retirement village setting, Bernard *et al.* (2007) explored the activities in this comparatively new innovation in the UK. Bae *et al.* (2021) suggested that ageing consumers tend to adopt fewer innovative products and services compared to younger age groups, although this finding is at odds with Nimrod's (2010, 2014) research. Buys and Miller (2007) examined retirement villages in Australia, where the residents reported greater levels of participation in leisure activities than non-village residents. In the case of care homes, Whyte and Fortune (2017) examined how natural leisure spaces stimulated expressions of individuality and provided opportunities to nurture relationships and interact with family members. In a French setting, Andriot and Roumilhac (2010) explored the travel experiences which care home residents enjoyed. Age UK (2014) outlined the growing range of housing options for an ageing population, which span shared ownership, retirement villages, charitable bodies, social renting, sheltered accommodation with 24-hour help and park homes (mobile homes), which are often installed on semi-permanent sites; altogether, this comprises around 730,000 units of specialist retirement housing (including 70,000 with care provided) in the UK.

The notion of active ageing promotes remaining independent in later life, which highlights the significance of people's living arrangements. In the UK, Torrington (2014) found that 93% of older people live in mainstream housing, 71% of which was owner-occupied and largely owned outright. In contrast, 5.6% lived in 'supported housing', with just under 500,000 in care homes. QoL studies point to the significance of one's home and local environment in later life (Pearce *et al.* 2006) as attachment to place is a key QoL enhancer or detractor given that cohesive neighbourhoods with positive social relationships reduce feelings of isolation, loneliness and depression in old age. Yet, as Torrington highlighted, most homes inhabited by older people are former family homes with multiple rooms and bedrooms and therefore too large for their needs. There is a mismatch between existing housing stock and the future needs of the ageing population as most single-occupancy properties are deemed to be too small. With housing such an important determinant of QoL, its interaction with leisure, tourism and the visitor economy is important within an ageing population.

Defining QoL, Bowling and Gabriel (2007: 828) suggested it 'theoretically encompasses a person's individual characteristics (e.g. physical and mental health, psycho-social well-being and functioning including feelings of independence and control over life) and external circumstances (e.g. socio-economic conditions, work, built environment and social capital)'. They demonstrated that having the freedom to do the things people want to do gives their lives 'quality' either inside or out of the home, and pleasure, based on satisfaction with life, mental harmony and social attachment. Social networks help the aged to engage in leisure to maintain independence. When frailty sets in and certain activities cease, they use a strategy of compensation to identify alternatives. Transition from independence to dependence, as health changes reduce functional (i.e. physical, mental and cognitive) abilities, requires changes to one's living arrangements. These may range from downsizing and moving house to the expanding range of housing options.

Yet there remains significant interest in how leisure and tourism participation changes when levels of independence and housing needs occur. Two key changes in relation to independent living may occur: enhanced care in-home to support growing issues of illness and disability (to retain independent in-home living) or admission to care homes or nursing homes. In each case, the compression of the time and spatial dimensions of leisure occurs, although Dening and Milne (2011) and Froggatt *et al.* (2009) take opposing views on participation levels in leisure in care homes. Froggatt *et al.* (2009) suggested that leisure enhanced QoL and promoted independence in care homes. Within care and nursing homes, most leisure activities are led by staff, with typical activities being watching television, listening to the radio, meeting other residents or visitors, reading, attending events and going for a walk. Less frequent outdoor visits away from the care home environment are an additional feature, for example to garden centres, the coast, parks and gardens, cafes and local events. Bradshaw *et al.* (2012) rejected the portrayal of care home life as 'sterile and devoid of meaningful experiences', as did Hall *et al.* (2011). Yet, Sprod *et al.* (2015) highlighted general concerns about the rise in sedentary behaviours among retirees, as well as a lack of understanding of this issue. Gibson *et al.* (2012) recognised that admission to long-term care homes requires significant adjustment and stressed leisure engagement's role in reducing isolation and loneliness. Altintas *et al.*'s (2018) exploration of nursing homes in France found that residents are typically admitted when their cognitive or physical deficits require continuous, 24-hour care, with adaptation often associated with levels of anxiety. Supervised and well-organised leisure activities may help the adaptation process and enhance perceived levels of independence, morale and overall well-being by giving new residents the option to participate.

The ageing market as a potential opportunity for seasonality

The ageing market offers a wide range of commercial opportunities for the visitor economy, dependent upon the health levels of the participants and their physical and mental capacity, as well as suitability and access to age-friendly infrastructure. In 2007, the WHO advocated the development of age-friendly cities/communities through a combination of bottom-up participation and top-down leadership in line with the life-course approach, followed by monitoring of the outcomes (WHO 2007b). Age-friendly initiatives offer opportunities for the visitor economy to address a perennial issue – seasonality. The ageing consumer is not impacted by institutional tourism (e.g. school holidays) and they have greater flexibility and responsiveness to price sensitivity and off-peak opportunities. In 2012, Age UK reported that the over-55s travelled more than any other age group in the UK, many of them by car. In the USA, McHugh and Mings (1991) studied the 'snowbird' flight of older people from cooler to warmer climates, while McHugh *et al.* (1995) explored the seasonal movement of retirees to Phoenix, Arizona, in recreational vehicles. A decade later, Smith and House (2006) suggested 800,000 over-55s travelled to Florida each winter, while 300,000 of the state's residents relocated in the peak tourism season. They also identified migration flows to Arizona, Massachusetts, Texas, Spain and Mexico, typically with visits lasting at least a month. Therefore, there is a clear scope for the visitor economy to tap into the flexibility of the leisure lives of an ageing population and to recognise where opportunities exist to utilise the capacity created in the low season.

Summary

This chapter has shown that we face substantial changes in the way ageing will affect the visitor economy in the future. The House of Lords (2019: 10) summarised this issue as follows: 'older generations face a society that is not prepared for their numbers or their needs as they age'. The same report also rightly criticised academic studies' tendency to treat older people as a homogeneous group, and to make generalisations that do not correspond to the lived experiences of many individuals. Scott and Gratton (2020) suggested that the projected increase in life expectancy (the House of Lords Select Committee referred to the '100-year life' concept) will result in older people remaining healthier (and therefore more able to travel) for longer.

That means we face the challenge of adapting the visitor economy in a way that addresses current and future barriers to participation by the ageing population. However, as the House of Lords pointed out, the elderly are not a homogeneous group. The leisure paradox means that we need to understand the intricacies and nuances of our ageing population so that, through minor changes to their behaviour and adaptations to the visitor economy, we will be able to nurture the new markets which active ageing and much higher proportions of older people will generate. Key to this will be making the visitor economy accessible and attractive to the older generation, as the next chapter will illustrate.

3 Ageing as a societal challenge

Visitor health, well-being, accessibility and the visitor economy

Introduction

The previous chapters highlighted a developing narrative around living well in older age, framed in terms of active ageing to remain physically and psychologically healthy, premised on the notion that more sedentary lifestyles emerge as we age. One aspect of this is the healthy ageing benefits offered by promoting recreational and tourist travel (see e.g. Morgan *et al.* 2015). This chapter examines how to encourage an ageing demographic to participate fully in civil society, and rejects the view of older people as a marginalised and isolated social group. We argue that the normalisation of ageing as a positive construct is vital to shape a visitor economy that is accessible for all. Drawing from an international study of visitor economy organisations, we examine the normalisation principle, which promotes positive experiences of ageing. We also examine accessibility as a concept and its implications for the visitor economy, introducing the principle of universal design and the role of the visitor economy in visitor well-being, together with the role of industry bodies in promoting these agendas. We conclude with an assessment of the policies and practices that are needed to develop a resilient age-friendly visitor economy.

Civil society and ageing

As Connell *et al.* (2017: 110) argued, 'globally, there is an increasing trend within both developed and developing countries to acknowledge the importance of how to achieve the objectives of a civil society, where the interests and needs of citizens/communities are met in a fair and harmonious manner'. Yet, as Chapter 2 emphasised, the ageing population is often portrayed as a Silent Generation or an invisible strand of society. The civil society concept seeks to embody the rights and participation of the whole population to achieve a more inclusive society. 'Civil society' has many meanings but, in its broadest sense, focuses on the community of citizens and the bonds that collectively identify and unite them. These different meanings have evolved from an esoteric and philosophical construct based on a predominantly European notion of citizenship that, as Ehrenberg (2013) identifies, has its roots in classical civilisation, with a focus on civility and the common

DOI: 10.4324/9781003039358-3

good. Civil society as an idea can be traced to Aristotle and discourses on political community, and through the Roman period, through the work of Cicero and the *societas civilis* – or a 'good society' for citizens. The concept saw a resurgence in the Age of Enlightenment that continued from the seventeenth to the nineteenth century, challenging the notion of the absolutist state, with its unelected representatives, and highlighting the importance of democracy (Hobsbawm 2010). The civil society concept was debated in key studies such as Ferguson's (1787) treatise, which examined the transition of society under industrial capitalism and the implications of this, a feature subsequently discussed by Marx and Gramsci.

In the post-modern era, civil society has become an in-vogue term that is widely used by politicians and policymakers. As Edwards's (2013) analysis of the term 'civil society' demonstrates, the challenge of theorisation has combined with the prevailing reductionist approaches and application by policymakers to denote its value as a 'big idea' for the twenty-first century. As the World Bank suggests (World Bank n.d.), civil society comprises 'a wide array of organizations: community groups, non-governmental organizations [NGOs], labour unions, indigenous groups, charitable organizations, faith-based organizations, professional associations, and foundations' (World Bank n.d.). This reflects the evolution of the concept from its initial development into what is now a highly contested term, harnessed by post-1945 liberalism to extol the virtues of the individual and people power, as embodied in Marshall's (1964) citizenship concept.

By the 1960s, thinking about civil society had shifted towards the rights of the individual in an attempt to understand how individuals and organisations could work together to fulfil their rights and obligations in society. Many of these debates were inherently philosophical, but as Birks (2016) illustrated, civil society concepts span a range of social, cultural, religious and political domains, including ageing. Probably the greatest impetus to the civil society concept was the emergence of ideologies that attempted to demonstrate how societal obligations towards the care of others, such as the ageing population, might be met. Giddens's (1998) ideas and other ideologies, including compassionate conservatism and new public sector management (McLaughlin *et al.* 2002), advocated filling the void left by the retreat of the state with civil society organisations. This encouragement of greater citizen involvement is not dissimilar to the rhetoric around active ageing. Giddens argued that the notion of citizenship was central to the development of a more active civil society to help with new local social policy initiatives. Many governments supported these philosophical shifts with greater resource allocations to the third sector (see Popple and Redmond 2000; Carmel and Harlock 2008) in the hope of achieving social policy objectives with regard to poverty and social care for the ageing population. However, critics argued that this represented a shift towards entrepreneurial solutions by harnessing the latent potential of business interests and skills to address grand social challenges, such as an ageing population. This was at a time when a decline in social capital and voluntary associations (Putnam 2000) was supposedly leading to a commensurate decline in voluntarism and community orientation, although this was primarily a North American phenomenon. Some theoretical explanations of these shifts are related to the repositioning of capitalism to recognise the business opportunities that are inherent in a third of the world's population being over 60 by 2050 (World Economic Forum 2015). Extending this idea, a concept that has challenged traditional thinking around business and society is the notion of *shared value*, formulated by Porter and Kramer, who argued that 'businesses acting as businesses, not as charitable donors, are powerful for addressing pressing issues we face' (Porter and Kramer 2011: 4). Shared value can be applied in three areas of business activity: 'by reconceiving products and markets; by redefining productivity in the value chain; [and] by enabling local cluster developments' (Porter and Kramer 2011: 5). This value can be enhanced by promoting recreational and tourist travel (Morgan *et al.* 2015). It is clear that ageing presents a compelling area for business and society to collaborate in in order to produce shared value.

Much of the disquiet with capitalism is linked to the way in which it creates inequalities and divides groups into 'haves' and 'have-nots' – issues that are reflected in ageing consumers' engagement with the visitor economy. Other concepts, such as *inclusive capitalism*, advanced by de Rothschild (www.inc-cap.com/participant/lady-lynn-forester-de-rothschild/), do not see any conflict between generating profit and meeting currently unmet needs, such as those of an ageing population. Such ideas mirror Porter and Kramer's argument, with its focus on managing businesses for long-term success, rather than short-term profit: the basic maxim is to make capitalism work for the masses. Benioff and Southwick's (2004) notion of *compassionate capitalism* advocates greater cooperation between business and civil society that goes beyond philanthropy and is anchored in serving the local community. Mackey's *conscious capitalism* (https://consciouscapitalism.org/), based on ethically grounded business dealings (Mackey and Sisodia 2013), focuses on stakeholder benefit rather than organisational short-term profit. In each of these models, the twin principles of capitalism – entrepreneurship and innovation – are harnessed for a broader purpose, beyond the simple profit motive, to contribute to societal development alongside economic development. In terms of the application to ageing, this represents a societal challenge for business, requiring them to rethink their purpose, values and contribution to the community in which they operate. Similarly, Jones's JUST Capital (see https://justcapital.com/) embodies many of these principles and offers a modified view of capitalism for the twenty-first century by prioritising civil society and fairness. Each of these models recognises that grand societal challenges, such as an ageing population, cannot be addressed in isolation by the state or the third sector without the cooperation of business; this applies in particular to the ageing population given the huge potential of the new market opportunities it presents.

Two key components of the civil society concept are that society must be free of ageism (Johnson *et al.* 2005) and that the often-silenced ageing population must be given a voice. Both elements involve the inclusion of older people in the community via activities, such as volunteering, and the promotion of active ageing as a social good (Kocka and Brauer 2010). A stable political system (Scott 2012) is a prerequisite for any civil society, and this will ensure that the state is able to counteract exclusion of people from tourism and leisure spaces. As Hall (1995) argued, one important principle of civil society is that it values each individual for their own contributions, rather than focusing on social division and subsequent labels.

The marketing literature posits that the ageing population comprises a largely untapped market that is ripe for development through value-creation processes, as evidenced in Barclays's (2015) report on hidden consumer spending amongst the ageing population. This report found that UK hospitality businesses derived 20% of their turnover from those aged over 65, yet only 5% of them saw this as an important market. Conversely, advocates of civil society reject segmentation approaches based on social difference and instead champion the *normalisation principle* by arguing that society should create a socially inclusive context to ensure that the ageing population is not marginalised. This situation is perhaps best reflected in the range of community and grass-roots clubs, societies and organisations that have been created, or are managed or supported by older people and which, in turn, benefit leisure resources and opportunities within the wider community. Furthermore, parts of the visitor economy such as visitor attractions and events managed by the state and voluntary sectors rely heavily on retired volunteers, as shown by studies of museums (Silverstein *et al.* 2001; Deery *et al.* 2011), theatres (Bernard *et al.* 2015), heritage railways (Goddin 2002; Rhoden *et al.* 2009), zoos (Fraser *et al.* 2009), and historic houses (Holmes 2003), to name but a few. This pattern is arguably less often reflected in commercial-sector leisure and tourism promotion and provision.

The normalisation principle debate

The normalisation principle can be dated back to the 1970s, when it was applied to people with disabilities. Its maxim is that we should not treat people differently from one another, regardless of any inherent constraints they may face. That is, we should not focus on differences, but rather treat everyone the same and fully integrate them in order to achieve a civil society. This idea is built upon the principle of people's right to participate fully in society, irrespective of their personal circumstances. Where barriers exist, we need to remove them. Normalisation has affected the design of services for people with disabilities, in parallel to deinstitutionalisation policies intended to facilitate supported living in the community.

Critics have challenged the normalisation principle in theoretical terms, but it is helpful for the ageing population in two ways. First, it challenges ageist attitudes to achieve widespread acceptance of old age. In Northern Ireland, the Office of the First and Deputy Minister (2004) indicated that progressing towards a more civil society will involve breaking down social exclusion to make the ageing population a normal part of society, which should help to overcome their isolation and marginalisation. Second, the normalisation principle helps to eradicate the division between 'haves' and 'have-nots' (i.e. those experiencing economic disadvantage) – a key element in creating a civil society and encouraging full participation, irrespective of age.

Some businesses have voluntarily adopted measures to progress towards normalising disability in society by viewing it as a market segment (i.e. *economic reasons*), to achieve corporate social responsibility goals (i.e. *altruistic reasons*) and/or in response to *regulatory measures* (e.g. disability legislation and the UK Equality Act 2010 as well as similar legislation around the world). The literature on the adoption of innovations (e.g. technology) and the business model literature (e.g. Amit and Zott 2012) suggest businesses adopt and engage with innovations or new ideas, such as normalisation, in distinct stages. But these models are very much business-focused and do not necessarily embrace the wider notion of moving towards a civil society and the challenge of developing age-friendly business experiences. One important model which helps to explain the normalisation principle is included in the World Health Organization's *Dementia: A Public Health Priority* report (WHO 2012), which focuses on integration of people with dementia in civil society, and the stages through which a society may pass en route to becoming dementia-friendly (DF). Figure 3.1, which is based on this model, offers a conceptual basis for exploring whether businesses in the visitor economy are aligned with the predicted phases of adoption and development of DF services and products (which, for our purposes, may be broadened to include age-friendly services and products). While any model is a simplification of reality, this one reflects the management and innovative capabilities of businesses and their recognition of the potential financial benefits of implementing change as well as other motivations, such as ethical business practice and corporate social responsibility (subject to which we will return in Chapters 6 and 7).

Despite this positive slant on how businesses and organisations might theoretically move towards a more dementia-friendly mode of operation, this process is not without challenges. Hollenstein (2004) described five potential barriers to the adoption of age-friendly practices among individual businesses (see Figure 3.2). Two key factors are resistance and a lack of motivation to adopt innovations when

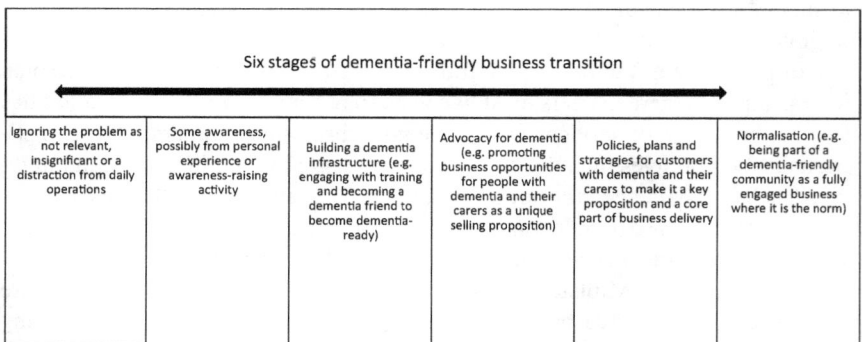

Six stages of dementia-friendly business transition					
Ignoring the problem as not relevant, insignificant or a distraction from daily operations	Some awareness, possibly from personal experience or awareness-raising activity	Building a dementia infrastructure (e.g. engaging with training and becoming a dementia friend to become dementia-ready)	Advocacy for dementia (e.g. promoting business opportunities for people with dementia and their carers as a unique selling proposition)	Policies, plans and strategies for customers with dementia and their carers to make it a key proposition and a core part of business delivery	Normalisation (e.g. being part of a dementia-friendly community as a fully engaged business where it is the norm)

Figure 3.1 The six stages of dementia-friendly business transition

Reprinted from Tourism Management, 61, J. Connell, S. J. Page, I. Sheriff and J. Hibbert, Business engagement in a civil society: Transitioning towards a dementia-friendly visitor economy, Page 114, © (2017), with permission from Elsevier

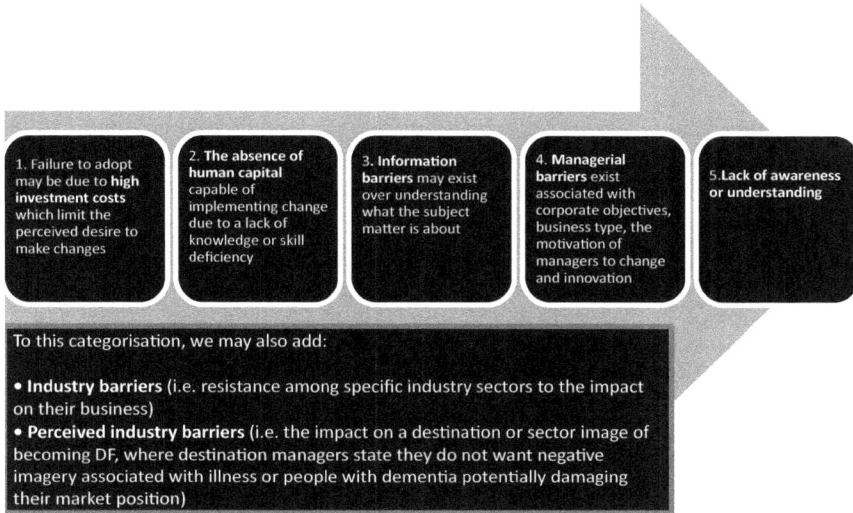

Figure 3.2 Hollenstein's (2004) five major barriers to adoption

Source: Reprinted from Tourism Management, 61, J. Connell, S. J. Page, I. Sheriff and J. Hibbert, Business engagement in a civil society: Transitioning towards a dementia-friendly visitor economy, Page 115, © (2017), with permission from Elsevier.

the barriers seem to exceed the perceived benefits. These often reflect the prevailing culture within the business, its leadership and/or its organisational predisposition to innovate. However, barriers may be overcome through intervention by an industry champion or an external body, such as a civil society organisation (CSO), which may set any business on the normalisation journey.

Civil society organisations

The definition of CSOs' remit has expanded over recent years. Cooper (2018: n.p.) described them as positioned in a space that exists outside the family, the market and the state:

> What constitutes civil society has developed and grown since the term first became popular in the 1980s and it now signifies a wide range of organised and organic groups including non-governmental organisations (NGOs), trade unions, social movements, grassroots organisations, online networks and communities, and faith groups ... Civil society organisations (CSOs), groups and networks vary by size, structure and platform ranging from international non-governmental organisations (e.g. Oxfam) and mass social movements (e.g. the Arab Spring) to small, local organisations (e.g. Coalition of Jakarta Residents Opposing Water Privatisation). Civil society has created positive social change in numerous places throughout the world.

Explaining the theoretical basis of community involvement, Etzioni (2000) employed the concept of mutuality. This concept depicts how people (and business leaders) develop an open-ended moral commitment within their local community, represented as a social obligation to that community. This obligation is one reason why businesses engage with grand societal issues such as ageing while seeking to create a culture of trust around shared values (see Birchall 2008). Self-interest may also explain why businesses engage in corporate social responsibility. Austin's (2000) explanation of mutuality described building common ground between business and the local community to illustrate where synergies exist for collaboration (see also Austin and Seitanidi 2012). Bowen *et al.*'s (2010) meta-analysis of business–community engagement identified three distinctive strategies that businesses may pursue as part of a socially responsible path: *transactional, transitional* and *transformational*.

From the community perspective, a great deal of progress within CSOs has been based upon a growth in voluntarism, where businesses' and individuals' altruistic (and other) motives, including participation by an ageing population, have prompted social challenges to be addressed via grassroots associations.

Kunreuther (2013: 55–6) defined grassroots associations as:

> a subset of the associations universe and in many ways they capture the ideal of civil society. These are groups where people come together voluntarily to advance a concern or interest, solve a problem, take an action or connect with each other based on something they share in common … [They are] characterised by democratic and less hierarchical forms of governance.

Such associations have a specific purpose – to mobilise citizens within a community, often around a specific issue. In terms of ageing, one of the principal concerns is enabling accessibility to ensure ageing people are able to fulfil their civil society ambitions.

The accessibility debate and ageing

Accessibility and ageing

Enabling everyone to participate, irrespective of age, is one of the cornerstones of achieving universal participation in a civil society. This issue is directly linked to accessibility. Hall (2020) argued that accessibility is necessary for full, competent citizenship, whereby reasonable adjustments are made for those with disabilities and other constraints created by ageing. Although the accessibility literature has focused on disability and the physical, social and psychological difficulties created by facilities and services that create barriers to use, the issue of accessibility is important for everyone in society, particularly older people, as specific constraints develop in older age around mobility. However, accessibility is not exclusively an issue for older people, and good design is becoming a more universally accepted approach that can benefit the presentation and ease of use of services.

The term 'accessibility' has been used in transport studies to examine mobility and people's ability to gain access to services, goods and different environments for leisure and tourism purposes (i.e. the visitor economy). Accessibility has been conceptualised in two ways:

- *As a positive element of human mobility* (i.e. physical access to participation in day-to-day activities); and
- *In a critical way when it is constrained* and thereby has an impact (positive or negative) on social welfare issues, including physical health, mental well-being and quality of life.

These two approaches to accessibility may be examined to a further level of granularity, as summarised by Gutiérrez and García-Palomares (2020: 407):

> Accessibility is a key concept in transport geography. It may be defined as the easy-to-reach desired destinations by means of a specific transport system. Accessibility is of great importance from the point of view of both regional development and social welfare. It depends to a great extent on the building of transport infrastructures, which in turn influences land use and mobility.

This quotation illustrates the importance of social welfare as an aspect of ageing where different environments create barriers to access, use and enjoyment. Disciplinary perspectives identify various approaches to these issues. Regional science, for example, has examined aggregate patterns of access within and across geographical space, using quantitative measures (Page 2021) to identify the scale and scope of issues that policymakers may need to address. Qualitative measures have been used to examine the social environment as well as policy issues rooted in political economy perspectives of ageing, which emerge as important determinants of accessibility. The political economy approach raises particular issues that are at the heart of society, such as the existence of structural barriers, as Chapter 2 illustrated with reference to 'haves' and 'have-nots'. But at a more operational level, we see that people are affected in specific environments and spaces by certain types of inaccessibility, highlighting the importance of looking at how spaces could be better designed to facilitate use and enjoyment. As Hall (2020: 7) suggested:

> One often cited solution to the problems of inaccessibility is so-called 'universal design', which seeks to produce an open, democratic set of spaces accessible to a diverse and dynamic society. Such design is concerned with not only the physical design of a space, but also with generating a social and embodied sense of inclusion. This change can be as simple as ensuring that everyone can enter a building through the same (front) door, and that internal layouts and communication technology are designed so that all can move around, and access people, services, and information. This democratic and inclusive approach to design seeks to include a wide range of people and bodies; reducing or removing barriers to access can lead to the empowerment of people with disabilities and others.

On this basis, changes to accessibility using the principle of universal design can enhance or improve the lived experiences of the ageing population, in a move towards a more normalised approach to thinking about the diversity of society that views it using a universal lens rather than focusing on age or disability as specific labels.

Universal design and accessibility

Hall (2020: 1) defined 'accessibility' as:

> The ability to move into and through an environment to reach places and use services. For people with mobility, sensory, and cognitive impairments, and those on the autism spectrum, there can be significant physical, attitudinal, and discriminatory barriers to access. Geographical studies have evidenced inaccessibility in the realms of urban and rural environments, transport systems, housing, health and social care, and technology. Geographers have also contributed significantly to conceptualizing accessibility in relation to disability, engaging with the social model of disability which emphasizes the political and structural production of inaccessibility.

One approach that seeks a more level playing field for accessibility is the concept of universal design (UD). As Darcy and Dickson (2009) observed, UD focuses on two specific issues that are relevant to the visitor economy: *physical infrastructure*, which allows access, as well as encompassing the products and experiences that are accessed as part of the visitor journey in a given environment; and the *touchpoints* that visitors have with businesses, organisations, services and the experiences they consume. Darcy and Dickson discuss these issues with reference to the 'continuous pathway' concept, which was developed as part of the Australian Standard for Access.

According to the Centre for Universal Design (http://universaldesign.ie/What-is-Universal-Design/):

> Universal design is the design and composition of an environment so that it can be accessed, understood and used to the greatest extent possible by all people regardless of their age, size, ability or disability. An environment (or any building, product, or service in that environment) should be designed to meet the needs of all people who wish to use it. This is not a special requirement, for the benefit of only a minority of the population. It is a fundamental condition of good design. If an environment is accessible, usable, convenient and a pleasure to use, everyone benefits. By considering the diverse needs and abilities of all throughout the design process, universal design creates products, services and environments that meet people's needs. Simply put, universal design is good design.

This declaration is based on the seven key principles that are outlined in Figure 3.3. As it suggests, the environments in which people interact should be designed to accommodate a diverse range of needs. UD is based on the principle of enhancing existing design to make it more inclusive, but the inherent design

Principle 1: Equitable Use (for everyone without segregating users)

Principle 2: Flexibility in Use (to accommodate a diversity of preferences and needs)

Principle 3: Simple and Intuitive Use (so that it does not require specialist knowledge by the user)

Principle 6: Low Physical Effort (allow use so the user maintains a neutral body position and a minimum of fatigue)

Principle 5: Tolerance for Error (to avoid unintended consequences and seeking to minimise hazards)

Principle 4: Perceptible Information (so that the design can be communicated easily regardless of the user's sensory abilities)

Principle 7: Size and Space for Approach and Use (ensure good usability without requiring further adaptations as well as an ability to accommodate adaptive devices)

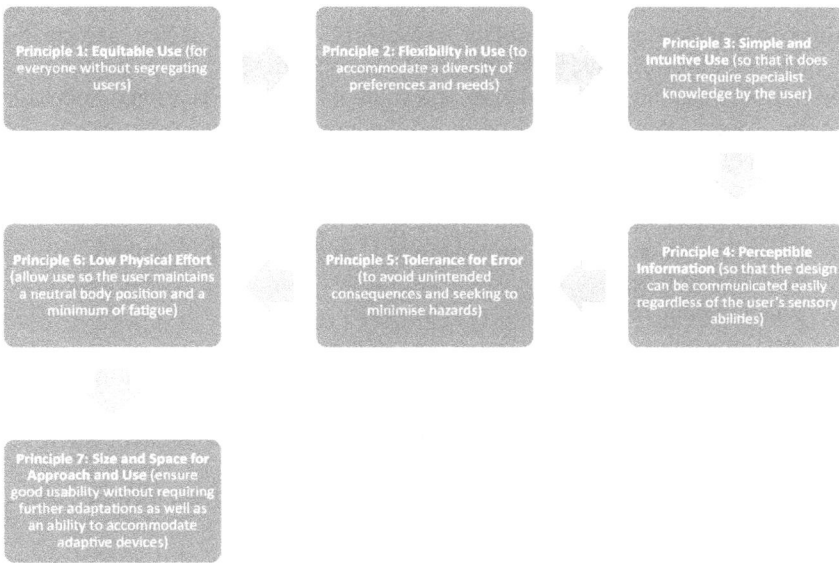

Figure 3.3 Principles of universal design
Source: Developed and simplified from http://universaldesign.ie

features should also be aesthetically pleasing. Therefore, it involves much more than simply meeting a range of accessibility requirements set out in legislative standards for short-term compliance needs. Any new design should make access easier for all users with physical, behavioural and other needs by building on almost 50 years of research into accessible design. Design-led accessibility is also informed by theoretical concepts, such as 'Rights to the City', initially developed by Lefebvre (1968) and Rawls (1971). These studies argued that policy should focus on how accessibility affects the less fortunate (Nazari Adli and Donovan 2018). However, the main stimulus to a more theoretical debate on accessibility and space arose with Lefebvre's later work (Lefebvre 1973, 1991), in which issues of justice and space were conditioned by politics, place and the policies that are developed and implemented for cities and other locations (see Harvey 2008). The main argument here is that space is not an empty void, but rather is 'filled with politics, ideology, and other forces shaping our lives' (Soja 2010: 19). So accessibility is a consequence of these forces' creation of unequal access and inequalities for individual people and groups, including the elderly. Consequently, citizens need to reassert their 'right to the city' through direct action (e.g. CSOs). In other words, achieving spatial justice for citizens requires collective action by the citizens themselves as a counterweight to capitalism, which targets certain groups at the expense of others.

UD is an important consideration for the visitor economy in relation to facilitating accessibility. The concept of accessibility is pivotal to the visitor economy and a core component of building a more inclusive society in terms of participation, as Finkel and Dashper (2020: 476) pointed out. As Finkel *et al.* (2019: 2) suggested, 'accessibility can be understood in terms of the measures put in place to address participation by those with impairments, both permanent and temporary, as well as both physical and mental, including perceived class and cultural barriers'. Focusing on one component of the visitor economy, Finkel and Dashper (2020) identified several ideas for event organisers, which have wider applicability to other settings. These include:

- Incorporating ramps and stairway lifts;
- Providing more space on pathways and between furniture to aid manoeuvrability;
- Widening doorways;
- Installing automatic doors;
- Providing more lighting, clearer signs, disabled toilets and other facilities;
- Utilising induction loops for people with hearing difficulties along with sign language interpreters and/or captioning; presenting information in alternative formats (e.g. Braille and audio platforms);
- Publicising information such as transport links, parking and facilities online.

Additionally, free attendance for carers and extra staff assistance is invaluable and is becoming an industry standard in visitor attractions.

Yet, implementation of these measures requires harmonisation of practices with the unique issues that affect the supply of visitor experiences to ageing visitors. Such harmonisation goes beyond the level of the individual business, visitor attraction or event, to encompass an integrated and seamless approach at destination level. Furthermore, accessibility is not just about *physical* access, as the increasing use of information communication technologies (ICTs) has raised interest in *internet* accessibility. As Arch and Abou-Zhara (2007) outlined, several age-related problems, such as visual impairment, hearing loss, deteriorating motor skills and cognitive decline affect the accessibility of the internet; they suggest that services and websites can be made more accessible by better design. This includes increasing contrast to make text stand out from backgrounds; easing navigation by clarifying links, augmented with simple images; and avoiding complicated drop-down menus and information overload (e.g. clutter caused by too many advertisements). In addition, sans-serif fonts in at least 12-point text size, and white, rather than coloured, backgrounds would increase readability (see Domínguez Vila *et al.* 2018, 2019, 2020).

The health and well-being of visitors and the visitor economy

There is considerable debate about the roles and responsibilities of the visitor economy towards visitors, much of which has focused on a growing body of work on the *duty of care* that businesses have to their clients, especially when the latter are travelling to unfamiliar environments or distant locations. Visitor economy businesses have tended to focus on their legal obligation to provide a safe environment for customers, which ensures a minimum standard of provision in relation to health and safety with a view to obviating common accidents and injuries (e.g. slips, trips and falls) and addressing visitor illness and unanticipated crises. As Chapters 4 and 5 exemplify, many of these issues may be amplified for the ageing traveller because the ageing process often creates additional difficulties in relation to accessibility. For example, various studies on accessibility and the visitor economy have explored: *visual impairment* (e.g. Lauría 2016; Kong and Loi 2017); *autism* (e.g. Hamed 2013; Dattolo and Luccio 2016); *accessible travel products* (e.g. Lyu 2017); *disability* (e.g. Dickson *et al.* 2017); *barriers in destinations* (e.g. Lee and King 2019); *flying experiences and disability* (e.g. Poria *et al.* 2009); *attraction and tourist site accessibility* (e.g. Israeli 2002; Mesquita and Carneiro 2016); *deaf and blind travellers* (e.g. Hersh 2016); and *accommodation accessibility* (e.g. Tutuncu and Lieberman 2016). For this reason, there is a growing expectation among ageing visitors that both international and domestic destinations will address the needs of a diverse customer base in the pre-visit information available, as well as providing orientation, information and customer support during the tourist experience. The same principles apply to the wider domestic visitor economy, where all leisure trips and activities should be inclusive, so providing information to the customer is vital to ensure artificial barriers to accessibility are not created unnecessarily.

The next section presents an analysis of destination management organisations (DMOs), their accessibility provision at the local level and the implications of this for the ageing traveller. This is followed by an overview of the information that different countries' national tourism organisations (NTOs) provide in relation to accessibility, specifically for the ageing visitor, to illustrate the developments that are already under way. First, though, to illustrate the scope and breadth of the current touchpoints within the visitor economy, Figure 3.4 shows how visitors to Scotland interact with a diverse range of businesses (both online and in person), as

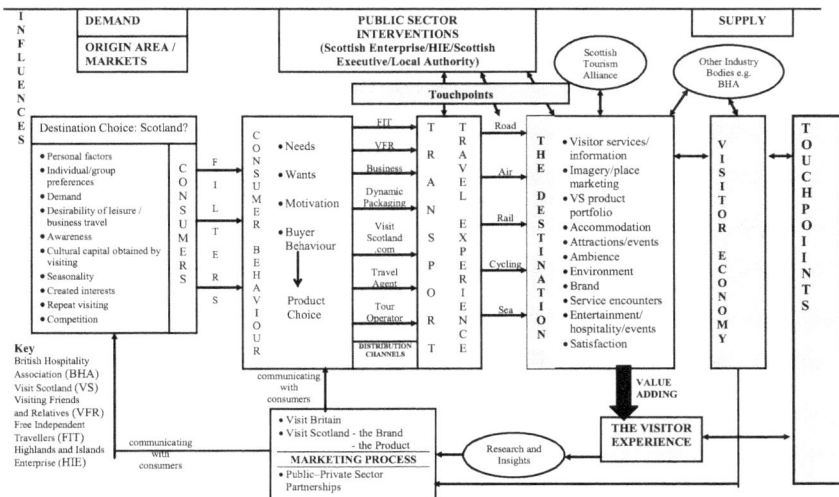

Figure 3.4 The scope of the Scottish visitor economy: touchpoints and potential accessibility issues

well as illustrating the environments they experience. This sheds light on some of the issues that may arise in relation to accessibility for ageing visitors. With these issues in mind, attention now turns to how the visitor economy has approached the issue of accessibility.

Accessibility, tourism and ageing: international and national perspectives

Roles and responsibilities: destination management organisations and national tourism organisations

The visitor economy, as Chapter 2 illustrated, comprises a diverse cluster of businesses and organisations focused on specific environments (i.e. destinations) which attract visitation. Leadership of the sector seeks to unify this fragmented group of businesses and organisations with a common focus – the visitor. Within destinations, leadership is typically organised at the national level through an NTO that markets and promotes the country. Meanwhile, DMOs manage the visitor economy at the regional or local level. There is an extensive literature on NTOs and DMOs (see, e.g., Connell and Page 2019b; Page and Connell 2020), which identifies the unifying feature of such organisations as being that they aim to promote their particular localities and manage diverse business interests, irrespective of their structure and source of funding. NTOs are typically funded by national governments to promote and manage tourism, but there are various other models, ranging from public–private partnerships to entirely industry-funded.

Destination management organisations

Dredge (2016) argued that a DMO's role is to stimulate growth, create value and support network development amongst stakeholders through a combination of market-enhancing and product-enhancing policies that are designed to address market failures. It is misplaced to assume DMOs will adopt a socially responsible approach when their underlying business models are focused on growth. Dredge (2016) suggested that they may be transformed from tools of industry to citizen service institutions as lo   ng as there is an element of public funding. However, at present, they tend to focus on generating income and engaging with stakeholders in collaboratively funded projects, as the state continues to withdraw from many areas of tourism policy. Although national DMOs are still largely government funded, there has been a considerable shift in emphasis over the last two decades so that many of them are now little more than marketing organisations (Pike and Page 2014), with marketing and advertising campaigns accounting for the majority of their spending. This has had an inevitable impact on their ability to perform their traditional role of managing tourism. Paddison and Walmesley (2018) discuss DMO adoption of new public sector management approaches, which has led to a perceived decline in local accountability in the cases observed.

Pearce's (2015) study of DMOs in a single country produced a wide-ranging typology (e.g. city/district council focused; regional tourism organisations; economic development agencies; and macro regional marketing alliances), to which national DMOs, with their strategic, country-wide overview, may be added. Pearce used the organisational behaviour literature to categorise the different types of DMO on the basis of their operational, marketing and inter-functional competencies (e.g. quality control) and general competencies (e.g. coordination). The organisational structure of DMOs varied from formal and tightly structured to loose and informal, but they all attempted to perform an enabling and enforcement role in relation to regulatory issues. However, the role of disruptive technologies must be recognised, and in particular, how the state contributes to that disruptive process when it withdraws existing funding and introduces a new model based on political ideology. For example, in England, the traditional structured model of provision was removed in 2010 and replaced with a framework that allowed new DMOs to emerge and form around new principles. This marked the start of a dramatic decline in public funding, which had fallen to £70 million per year by 2015, including business support services provided by local government bodies (www.gov.uk/government/collections/local-authority-revenue-expenditure-and-financing). The new structure initially focused on closing England's regional development agencies – economic development-focused bodies that had funded many DMOs – and facilitating the transition to local enterprise partnerships (LEPs). A total of 38 LEPs have been formed since 2010, funded by two government departments. However, these are not the sole conduits for DMO development in the new funding landscape, as many other organisations have emerged that are driven by commercial reality and local action, rather than government policy. Some of the latter are spatial- or resource-focused state entities (e.g. National Parks), while London has a specific body reflecting its world city status (London and Partners). They have wide remits that may encompass management as well as marketing and promotion. The rationale of these new DMOs is to coordinate stakeholders, fulfilling the higher-level competency identified by Pearce (2015), with a much stronger focus on tourism businesses, the community and other associated interests (e.g. transport operators). In management terms, Visit England is the lead organisation within the new DMO landscape, which also includes public-sector bodies funded by local authorities, organisations focused on defined boundaries, private companies, community interest companies, public–private partnerships and pan-geographical organisations. It has had responsibility for creating a knowledge base that will enable the new generation of DMOs to meet the requirements of the Equality Act 2010 and ensure that destinations take reasonable steps to allow all visitors to access destinations. As later chapters explore, these steps include the production of a range of guides that focus on operational issues to disseminate best practice and guidance on how to make destinations more accessible. One issue this raises, of course, is the question of how knowledge is produced and the discourse(s) it represents. Are people with lived experience at the heart of knowledge production, or are these simple checklists to help businesses make 'quick fixes' that do not necessarily address the issues in a meaningful way but help to meet legal requirements? This debate underpins much critical social research on disability, health and ageing studies.

We now focus on two surveys – one of NTOs and the other of UK DMOs – to gauge the progress the visitor economy has made on the issue of accessibility, specifically in relation to the requirements of the ageing population.

National tourism organisations

Around 180 countries (the number alters slightly on an annual basis) are members of the world's main tourism body – the United Nations World Tourism Organisation (UNWTO) – which means that only around 36 are not. Many of the members have an NTO as a standalone body, part of a government department or in some other organisational form. Table 3.1 presents all of the official statements on accessibility that appeared on the websites of the top 25 NTOs (in terms of visitor numbers) in 2020 as well as any comments they had to make about provision for older visitors. It reveals that the majority of these NTOs had little to say about ageing travellers' access needs; as a result, many potential visitors would have been obliged to search elsewhere for information.

Table 3.1 Examples of leading tourism destinations' and national tourism organisations' accessibility statements and provisions for ageing travellers

National Tourism Organisation	Accessibility Statement/Comments from National Tourism Organisation website	Mention of elderly travellers either as market segment or in relation to accessibility
Explore France	'Making tourist sites fully accessible is essential for all French destinations today. The national labels "Tourisme & Handicap" and "Destination pour tous" provide information on accessibility in France and where it's provided. Many tourist associations also have regional or local initiatives in place to help.'	
Espania (Spain)	'There are numerous monuments, museums, nature areas, accommodation options and restaurants offering services, settings and activities that are partially or	

(Continued)

Table 3.1 (Continued)

National Tourism Organisation	Accessibility Statement/Comments from National Tourism Organisation website	Mention of elderly travellers either as market segment or in relation to accessibility
	totally accessible.Spain is becoming increasingly prepared for accessible tourism – tourism for everyone. Indeed, enormous efforts are being made disabilities by eliminating barriers, providing access to cultural and natural resources, creating standards for transport, building and urban planning.'www.spain.info/en/	
China National Tourism Office	'The barrier-free environment in China is developing fast, especially in cities like Beijing, Shanghai, Tianjin, Guangzhou, Shenzhen, Shenyang and Qingdao.'www.travelchina.org.cn/	'For senior tour or travel with senior or disabled travellers in China, recommended destinations will be big cities or hot tourism cities in China with convenient public facilities and friendly locals.' Link to recommendations and ratings of locations for senior travellers:www.topchinatravel.com/customer-center/destinations-for-china-travel-with-seniors-disableds.html
VisitBritain (Intern ational) Visit England(Dome stic)	VisitBritain: 'The facilities on offer for visitors with special needs are steadily improving. Recently designed or newly renovated buildings and public spaces that provide lifts and ramps for wheelchair access are becoming more common. It is important that you make your needs very clear when booking any service or facility in the UK. Your impairment may not be obvious to other people, and it's best not to assume that reservations staff will know your needs. Buses are also becoming increasingly accessible, and, if given advance notice, train, ferry or bus staff will happily help any disabled passengers. Ask a travel agent about the Disabled Persons Railcard, which entitles you to discounted rail fares. As well as this, many banks, theatres and museums now provide aids for the visually or hearing-impaired.	

(Continued)

Table 3.1 (Continued)

National Tourism Organisation	Accessibility Statement/Comments from National Tourism Organisation website	Mention of elderly travellers either as market segment or in relation to accessibility
	Specialist tour operators, such as Tourism for All, cater for physically disabled visitors.' VisitEngland: 'England's lushly beautiful countryside, vibrant cities and dramatic coastline are there for everyone to enjoy, including disabled travellers and those with access needs. There's plenty of information and guidance on our site as to the best places to visit, where to stay, and how to get there – from assistance at railway stations, to accommodation that has been specifically assessed for those with access needs.'www.visitbritain. com/gb/en	
Visit the USA	No mention of accessibility.	
Go Turkey	'The resorts which are located in relatively flat areas and are, therefore, better suited to wheelchair users are: Marmaris, Icmeler, Dalyan, Fethiye, Calis Beach, Side, Kemer. Anyone who has difficulty in walking should certainly avoid resorts on steep hills such as Kalkan and Turunc. Obviously, hotel locations vary so do check before booking. 438 disabled-friendly blue flag destinations …100 of these destinations fully equipped, with accessible sunbeds, umbrellas and sea access.'www.goturkiye.com/	Picture on webpage on disability access features senior citizens.
Italia Agenzia Nazionale Turisimo(Italy)	'Over the course of the last century tourism has become a primary social need. Not only is it an extraordinarily important economic factor, but it is also a means to increase knowledge and personal development. For all these reasons, today it is essential to grant access to tourist experiences to all citizens, regardless of their personal, social, economic and any other condition that could limit such experiences.' www.enit.it	Senior citizens are mentioned as part of the general statement of accessibility in relation to 'Senior Citizens, Persons with Disabilities, Nutritional Allergies and Special Economic Conditions'
Visit Mexico	No mention of accessibility.	
German National Tourism Board		Ageing and retired tourists are mentioned: 'Germany offers a wide range of

(*Continued*)

Table 3.1 (Continued)

National Tourism Organisation	Accessibility Statement/Comments from National Tourism Organisation website	Mention of elderly travellers either as market segment or in relation to accessibility
		activities for people of a certain age …Germany not only offers people "in the prime of their life" a huge range of leisure activities, the country's infrastructure also makes travelling easy, whether you're alone, a couple or part of a group. Accessibility for disabled people, medical care and excellent transport links, including bus and rail services, are some of Germany's special features.'www.germany.travel/en/travel-information/50-travellers/senior-citizens50.html
Amazing Thailand	'Thailand is not an easy place to visit for people with reduced mobility or other physical challenges. The larger resorts and tourist attractions provide facilities for disabled people, but in rural areas public transport is limited and often inaccessible to wheelchair users. Moving around the city can be extremely difficult for disabled people. The streets and pavements are uneven and few buildings provide ramps and handrails to aid disabled access. Guide dogs are rare and there are few audio signals for the blind at traffic crossings. Nonetheless, a project has been announced by the Bangkok governor along with the Disabled People International Asia-Pacific Region to ensure that Bangkok pavements are easily navigable for those with reduced mobility. A commitment has also been made to make public transport more accessible. Public transport is not usually equipped to facilitate disabled access. Public buses are inaccessible to wheelchair users. Disabled people are usually forced to travel through the cities by taxi. However, few taxi drivers	

(*Continued*)

Table 3.1 (Continued)

National Tourism Organisation	Accessibility Statement/Comments from National Tourism Organisation website	Mention of elderly travellers either as market segment or in relation to accessibility
	are experienced or trained in helping a wheelchair-bound customer into and out of their cars.'www. tourismthailand.org/Articles/even-more-amazing.	
Visit Dubai	'Dubai strives to welcome all visitors, including wheelchair users, those with sensory impairments and senior travellers. A national strategy and special advisory board was launched in 2017 to further empower services for people with disabilities, known as "people of determination" across the UAE, including travellers and tourists Follow our handy guide on accessibility in the city. Several attractions across Dubai ensure easy accessibility. Public parks also have designated parking spaces, complementary wheelchairs, and staff who are trained in sign language. There is also a comprehensive guide to the city's public parks available in Braille for the visually impaired, and disabled-access washrooms at most major attractions. People of determination benefit from free entry or discounted tickets at a number of key attractions across Dubai, including public parks, waterparks and theme parks.'www.visitdubai.com/en/	'Senior travellers' are mentioned in the accessibility statement.
Egyptian Tourism Authority	No mention of accessibility.	
South African Tourism	'Universal access refers to the ability of all people tohave equal opportunity and access to a service orproduct from which they can benefit, regardless oftheir social class, ethnicity, ancestry or physicaldisabilities.' Link to: www.southafrica.net	Broad, all-embracing approach to accessibility.
Kenya Tourism Board	No mention of accessibility.	
Visit Morocco	No mention of accessibility.	
Argentina Tourism	No mention of accessibility.	
Visit Brasil	No mention on the official NTO website, but a number of destinations are listed	

(*Continued*)

Table 3.1 (Continued)

National Tourism Organisation	Accessibility Statement/Comments from National Tourism Organisation website	Mention of elderly travellers either as market segment or in relation to accessibility
	on a different website as suitable for people with disabilities or reduced mobility (Sacorro, Bonito, Brotas, Fernando de Noronto, Fortalez, Foz do Iguacu, Maccio and Rio de Janeiro).	
The Islands of the Bahamas	'The Ministry of Tourism is aiming to tap into a billion-dollar travel sector that has long been overlooked and will now begin to place accessibility at the heart of tourism development in The Bahamas.'	
Cuba	No mention of accessibility.	
Go Dominican Republic	No mention of accessibility.	
Visit Jamaica	No mention of accessibility.	
IndiaTourist	No mention on the official NTO website, but the Ministry of Tourism's guidelines for making tourism barrier free were launched in 2020.	
Tourism Australia	'Accessible tourism is the ongoing endeavour to ensure tourist destinations, products and services are accessible to all people, regardless of their physical limitations, disabilities or age. People with access requirements include those with young children in prams, seniors with mobility requirements and people with permanent or temporary disabilities. Their access requirements may include: Physical/mobility; Hearing; Vision; Cognitive.'www.australia.com	'Seniors with mobility requirements' are mentioned in the accessibility statement.
Destination Canada (2018)	'For details about accessible transportation in Canada and links to resources for travellers with special needs, visit the Access to Travel website.' Link to:www.canadianaffair.com	
Visit Greece	No mention of accessibility.	

Source: Authors' analysis of 25 NTO websites, July 2020. Unless otherwise stated, this information was accessible at the time of going to press.

In France, the main initiative, 'Tourisme & Handicap', which was launched in 2001,

> has the intention of giving people accurate, objective and consistent information about the accessibility of tourist sites and facilities in regards to the 4

handicap types (motor, mental, auditory and visual) and to develop an adapted tourist offer ... Even though establishments haven't always been accessible in France for those with disabilities, making tourist destinations accessible for all is now a definite priority for French cities, towns and regions.

Explore France website

In Spain, the focus has been on developing 'more and more initiatives ... that guarantee a cultural tourism which is accessible to all. Adapted routes, specialised services, adapted facilities and accessible guides are now a reality in Spain' (Espania website).

However, in 2020, the market leader was undoubtedly the UK, led by VisitEngland, with its multiple accessibility guides for different sectors and its promotion of accessibility initiatives:

Activity breaks are another great option, and again there are plenty of facilities for disabled travellers. Wheels for All offers adapted cycles at many centres including Cumbria, Sheffield and Northumberland, while fishermen can spend a lazy day on the water courtesy of Wheelyboats, which offer independent boating for wheelchair users at locations from Redruth to Ripon ... For families, the spectacular aquarium at The Deep in Hull is ideal for those with sensory needs; particularly on a 'quiet day' when lighting and sound levels are adjusted, BSL-signed presentations are available through the day and there are multi-sensory and interactive experiences.

VisitEngland website

In marked contrast, even though it comes just below the UK in the list of the world's most visited destinations, Turkey's NTO simply states:

Sorry but many Turkish resorts and cities are not planned good enough for wheelchair access, which can make life difficult ... [H]owever, you will find that Turkish people always try their best to be helpful and will gladly improvise to find a solution.

Go Turkey website

DMOs in the UK

Connell and Page (2019b) studied a range of DMO websites to explore the infrastructure that is currently provided to accommodate access needs. One area they sought to focus on was the level of accessibility at destination touchpoints. This analysis confirmed Ancient and Good's (2014) argument that two key features are essential when developing destinations for people with access needs: *personalisation* (to increase accessibility and usability) and *user acceptance of the technology* (see below for detailed information on this survey). Gallistl and Nimrod (2020) identified four categories of older users of digital leisure: innovative traditionalists; entertainment seekers; selective content consumers; and eclectic media users. Yachin and Nimrod's (2021: 2) study pointed out that the 'Innovation Theory of Successful

Aging' (Nimrod and Kleiber 2007) 'may promote growth, productivity, independence, and a greater sense of meaning in life' where specific goals are fulfilled. For example, Hauk *et al.* (2018) found that one of the key motives associated with technology adoption is the perceived importance of its usefulness, as is evident in Yachin and Nimrod's (2021) analysis of grandmothers' use of Facebook for leisure purposes. The latter illustrated constrained use as a form of restrained innovation that demonstrated engagement with technology among the older women in the study. That said, in primary data collection based on semi-structured interviews, we encountered a wide range of views on this issue. For instance, one of our visitor economy respondents (R2) outlined not only their own business's use of technology but the broad combination of communication channels that they employed to target ageing consumers:

> Well, growing numbers of people that age fall into a category which I'm sure you've heard of – silver surfers – who have endorsed (a) the internet and (b) certain forms of social media … We send messages out via Facebook, and Facebook is quite a strong one in terms of getting messages through to people … and also via Twitter and … on Instagram … So we communicate with them by a variety of means, but a lot of the traditional means of communication, with some PR techniques, which is my background, in newspapers, radio and television, they're very, very strong indeed … So we try and communicate by whatever means we can and that's also down to little things like parish magazines or interest magazines in and around the county. So we try and send stuff to as many outlets as possible, as well as national magazines.

Meanwhile, a representative of an age-related charity working in a large conurbation (R15) stated that we need a balanced view of technology because

> there is a big digital divide which we've tried to get the … [public-sector organisation] … and other public bodies to recognise … There's a large number of people who you can't reach through the internet, you can't reach by email, and therefore you need to be doing more printed material and to do phone calls, or at least have phone lines that work for people to access information, and possibly through radio. We have done some work with community radio stations to address issues like that. But we have a big problem with Covid now, which is that we really can only communicate by phone, online, through the website, by email and text … I would say half of the people that we get who are digitally excluded are digitally excluded because of money. They know how to use a phone, they know how to use their computer, but they just don't have the money to get an upgrade or they don't have the money to get a decent contract that will allow them to access it.

This has an important bearing upon how the ageing visitor uses websites for trip planning, as this is clearly contingent upon having access to the necessary technology, so those with low household incomes may be excluded.

To gauge UK DMOs' degree of engagement with accessibility issues, Connell and Page (2019b) analysed the websites of 127 fully operational destination management organisations, all of which were VisitEngland-recognised. Key content extracted from these websites included: extent of content relating to accessibility; range of visitor services identified as accessible; evidence of destination-level promotion; and external links to further information, for example on the funding of the DMO and its organisational status.

Overall, 65% of DMOs had information on accessibility on their websites (excluding the legal obligation in terms of the accessibility of the website itself). The DMOs that performed well tended to be limited companies and Local Authority-funded DMOs. One might expect all DMOs to provide such information, given the passing of the Equality Act 2010, but some of them were still in the process of developing this facility, as was pointed out by 33 of the organisations in their accessibility statements. The most comprehensive example was a government-funded organisation that incorporated a guide and audio-visual content. Other DMOs outlined the scope of their provision throughout the visitor journey, as well as known issues such as difficulties that may be encountered in historic buildings and infrastructure that cannot be adapted to meet some visitors' needs. Among the most ambitious DMOs was one that announced its intention to make the destination accessible to every potential visitor.

These accessibility statements help us to begin to form a picture of what best practice to make destinations more accessible might look like, but there is some way to go, as only 22 DMOs (just 17% of the sample) had a separate webpage devoted to information on accessibility issues, and only 7% offered a specific guide on accessibility. Potential visitors to other destinations could typically be obliged to perform online searches for information on particular aspects of accessibility. Some 50% of the DMOs – primarily those that were run by local authorities – provided information on accessible accommodation on their websites, which probably reflects the local government tradition of creating visitor guides for distribution by hotels and guest houses, attractions and tourist information offices. Similarly, 55% of the DMOs – with the largest proportion again being local-authority run – listed searchable, accessible attractions. Surprisingly, accessible events were listed on only 34% of the websites, although 64% of the DMOs that were operating as limited companies publicised such activities, as did those funded through Business Improvement Districts (BIDs), given the business opportunities afforded by an extensive events programme. Supporting infrastructure (e.g. accessible toilets and opportunities for mobility-assisted shopping) was evident on 45% of all websites, dominated by the limited companies (58%), while only 40% of all websites listed accessible transport options. Again, limited companies performed best in the latter metric (58%), although they were worst when it came to listing hospitality services (just 21%). Only seven of the websites referred to the National Accessible Scheme (NAS) operated by Visit-Britain.org. This scheme rates the accessibility of accommodation and provides cases studies and guides for businesses. For example, simple adaptations to the design and layout of rooms, such as providing handrails, step-free areas, wider areas around beds and more space in dining areas, may make a significant difference. Just one DMO actively promoted this scheme on its website. Several other websites referred to the Equality Act 2010 and businesses' obligations in relation to it (e.g. the Equality Act replaced and subsumed the Disability Discrimination Acts 1995 and 2005) in their members' sections. Just one outlined the scope of the Act and highlighted the need to treat everyone fairly, irrespective of age, gender, race, sexual orientation, disability, gender reassignment, religion or belief. Clearly, then, UK DMOs are still a long way from achieving the two principles outlined by Ancient and Good (2014). With these issues in mind, we now turn to building the case for enhancing accessibility in the visitor economy.

Building the case for enhancing accessibility: a role for best practice?

There is a growing debate about the efficacy of using legislation to address accessibility issues. While it may provide the framework and broad principles within which society is expected to operate, the data presented above on NTOs and DMOs suggests that it is insufficient in itself to achieve success. In the absence of a rigid (and expensive) audit regime, other tools and techniques might need to run in parallel with legislation. One such tool, which may be championed by a specific group or organisation in the visitor economy (such as a DMO or an NTO), is a best practice guide that showcases how services and experiences can be made more accessible, especially for ageing visitors, through simple changes to business practice. Yet, businesses that engage with ageing people in their day-to-day lives and leisure time face a range of challenges in addition to physical accessibility issues:

- Recruitment of an agile workforce that can meet the needs of *all* 'ageing' people, from the highly active to the inactive and sedentary.
- Handling death when it occurs on the premises, or the after-effects of the event.
- Managing a volunteer workforce who may need training with regard to ageing issues.
- Maintaining empathy when dealing with visitors who make immediate and urgent demands for service provision.
- Accommodating the very diverse market known as the ageing population of visitors, which includes those aged from 55 to over 100.
- The digital exclusion of some sections of the ageing population.
- Overcoming various barriers to participation.

Yet even where studies exist that promote best practice examples (e.g. Buhalis *et al.* 2012), based on the concepts of accessibility (e.g. Buhalis and Darcy 2011), establishing best practice remains conceptually weak in tourism, and its meaning is often ill-defined. These academic studies remain distant from the very audiences they need to influence as the visitor economy has a low take-up rate of academic research. This low take-up is closely related to visitor economy businesses' absorptive capacity, which is their ability to discover new research information and to then assess it, digest it and then commercialise it to improve their performance, typically by innovating. Many of these visitor economy businesses have low levels of innovation and so their capacity to reach out to new markets, such as the ageing visitor, may be very limited if left to those individual businesses. Given the very home-based leisure activities of ageing people, visitor economy businesses will need to pursue more innovative commercial strategies, potentially through collaboration and cooperation (e.g. as clusters of businesses) in order to make a step change in their innovative capacity and desire to reach out to this market. We will return to the subject of best practice in Chapter 6, where we explore its successful implementation. Even so, as Avramov *et al.* (2003) point out, the home-centred and relatively inactive leisure lifestyles of retirees after ceasing paid work in Europe pose a challenge for expanding demand.

Summary

This chapter has argued that, in order for an ageing population to become active citizens in a civil society, we need to look at how we theorise and understand the changes that need to be made at the societal level and, more specifically, in the visitor economy, to enhance accessibility. Research has an important role to play here, as it allows us to understand the extent to which accessibility principles are already being pursued. Nevertheless, accessibility could be better conceptualised as a more integrative concept for the visitor economy and linked to the normalisation principle, which asserts that full accessibility is the ultimate goal. While legislation and compliance provide a framework for the functioning of the visitor economy, enhancing visitor experiences and encouraging more participation means motivating and creating a higher level of demand. Stimulating such demand needs to be accompanied by an improved – and increasingly digital – communication of the greater accessibility (based upon principles of UD, wherever possible) within the sector to inform, reassure and influence awareness, interest and decision-making.

This chapter has demonstrated that the visitor economy still has a long way to go in terms of addressing accessibility issues, although industry champions may be able to show businesses how to build new market opportunities, particularly by referring them to best practice examples. Yet, building these new opportunities will also require detailed knowledge of ageing consumers and the barriers they face as well as the myriad specific needs that the visitor economy must recognise and understand. These issues are the focus of the next two chapters.

4 Accommodating visitors with specific needs

Perspectives on visual, auditory and learning needs

Introduction

One of the central tenets of accessibility and the concept of universal design (UD) is that anyone with a specific need should be able to access the full range of resources, experiences and locales in the visitor economy, as Chapter 3 highlighted. Lee *et al.* (2020) examined major increases in life expectancy and established that the over-60s are one of the fastest-growing age groups, already comprising 13% of the world's population and 25% of Europe's population. The challenge for the visitor economy is how to meet the broad spectrum of diverse needs presented by ageing visitors. Alternative discourses within critical gerontology offer some interesting approaches to theorise and explain why and how that might be achieved. Critical gerontology, with its focus on social justice and the promotion of access, identifies how society shapes ageing. Wood *et al.* (2018) demonstrated that studies such as Tornstam (2011) tend to divide the literature on critical gerontology into two domains: (a) *the misery perspective*, which focuses on the difficulties posed by ageing, such as mobility, loneliness and isolation; and (b) *the positive approach*, in which retirement is not viewed as a trauma but as an opportunity for rejuvenation, the latter of which was covered in Chapter 2. In this chapter, and throughout the book, we adopt the positive approach towards ageing and explore how the visitor economy can make changes that make visitor experiences accessible and rewarding for older people, in a discussion informed by different disciplinary perspectives, including critical gerontology.

Our starting point is to ensure accessibility is embedded in the visitor-journey concept. Lane's (2007) analysis of the visitor journey in London identified two principal components relevant to tourism and leisure: (a) *travel to the destination and accessibility*, and (b) *mobility within the destination*. We advocate the use of touchpoints as a methodology with which to understand accessibility; it focuses on key interactions, and how visitor mobility and flow connects the various elements of service provision in the visitor experience. This approach consists of two fundamental elements:

- *Travel to the destination*. Whether by private or public transport mode, short or long haul, or even undertaken with travel itself as the focus of a

DOI: 10.4324/9781003039358-4

holiday, such as an extended tour or cruise, the travel component of tourism and leisure consists of multiple elements. Darcy (2012b) examined this in relation to air travel and found that it could be broken down into the following touchpoints: pre-travel planning; boarding and disembarking; seat allocation; on-board personal care issues; equipment handling; and customer service. At each stage, Darcy outlined some of the experiences and anxiety that disabled travellers may encounter, and which may be amplified by age.

- *Travel within the destination*, or mobility for local residents' leisure day trips from home to the local area or region, which make use of elements of the visitor economy.

The term 'the destination' is a universally accepted construct which Morrison (2019: 4) defined as a 'geographic area that attracts visitors', but it can be subdivided into a number of components. Page and Connell (2020: 327) described these as the six 'As':

- Available packages;
- Accessibility;
- Attractions;
- Amenities;
- Activities; and
- Ancillary services.

In mature destinations, a public, public–private or private-sector organisation may be responsible for the coordination, planning and promotion of the destination, as outlined in Chapter 3. Although mainly evident in tourism, this concept also applies to leisure, as Figure 4.1 shows, where the focus is on leisure resources with which people engage outside the home. Frequently, the same recreation and environmental resources and opportunities are promoted to and used by tourists and leisure visitors as elements of destination supply, with some divergence around community resources.

The critical component here is accessibility in the visitor economy, which generally means the visitor's mobility or transport to the destination and within it. However, for the purposes of this book, the term has far greater scope, encompassing application of the principles of UD in relation to ageing consumers as well as mobility and access within the destination. Therefore, this chapter will examine these broader principles of accessibility for ageing consumers in terms of mobility and access for those with impairments or impediments that individually or collectively combine to cause multiple conditions that could hinder that access. While it is outside of the scope of this book to examine all of the medical and health conditions that may affect ageing consumers, recognition must be given to the coalescence of conditions, known as co-morbidity, that create complex situations for visitors and about which visitor economy businesses need to be aware. Using labels for people with medical conditions is not necessarily an inclusive way to understand each

Frequency of Use

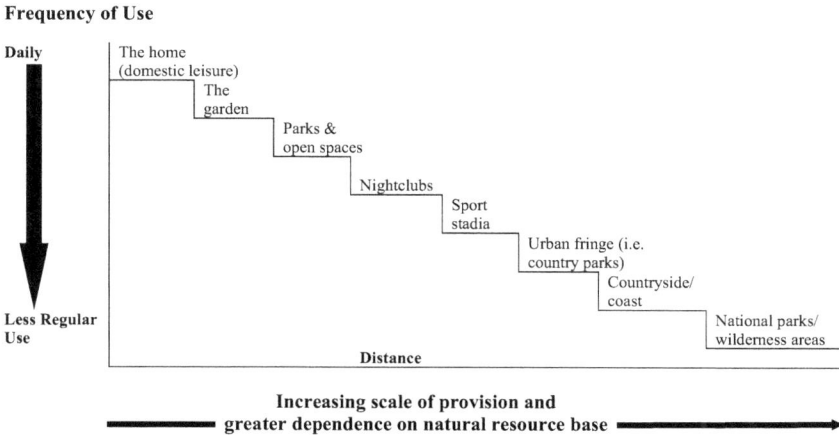

Figure 4.1 A continuum of leisure resources

Source: *Page, S; Connell, J; Leisure: An Introduction, 1st Edition,* © 2010. *Reprinted by permission of Pearson Education Limited.*

individual but, from a practitioner perspective, looking at single conditions that can be termed 'a disability' is one approach that has gained acceptance. This is a useful way to understand the types of issues people face in accessing the visitor economy, as we explore in this and the subsequent chapter. However, prior to that discussion we examine the demand (i.e. the people who participate) for tourism and leisure activity globally, in order to understand the scale and nature of consumer demand.

Global travel demand and ageing: travel behaviour and trends

In 1991, the international tourism industry employed 112 million people worldwide and generated over US$2.5 trillion at 1989 prices. By 2006, this had reached 234.3 million people employed; in 2012 the figure had grown to 260 million jobs, and in 2019 the industry generated US$2.9 trillion for the global economy. According to the World Travel and Tourism Council (WTTC; www.wttc.org), in 2019, travel and tourism made a US$8.9 trillion contribution to global GDP (10.3% of the total) and employed 330 million people (around 10% of all employment). While Covid-19 has reversed these growth trajectories, the visitor economy has a long history of recovering well after previous interruptions to demand (including the global flu pandemic of 1918–19). Therefore, it is not surprising that the WTTC continues to argue that tourism is the world's largest industry, especially if one includes the broader visitor economy.

The OECD's *OECD Trends and Policies* (OECD 2018) provides a useful framework in which to understand the influences on current and future patterns

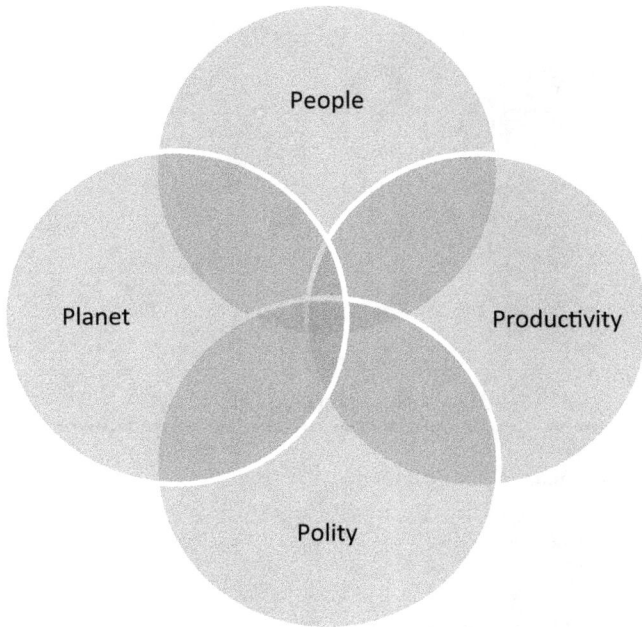

Figure 4.2 Megatrends shaping the future of tourism

of tourist travel and the wider visitor economy in relation to four interconnected global megatrends (see Figure 4.2):

- *People*, specifically, the implications of an ageing population and a growing middle class on visitor demand.
- *The planet*, with a focus on climate change and its impact on resources such as food, energy and water, all of which can be adversely impacted as a result of visitor demand, through consumptive tourist use and tourism development, especially in fragile environments (e.g. island microstates).
- *Productivity*, in terms of the growth of new technologies that have created the sharing economy (e.g. Airbnb), and the potential impact of innovations such as automation and robots, as well as artificial intelligence, which may affect interactions and touchpoints in the visitor experience.
- *Polity*, that is, the role of government and its impact on future patterns of mobility related to issues in the other three spheres, such as climate change, the ongoing commitment to uninhibited global travel and a reluctance to manage demand, despite resource depletion.

These four megatrends collectively interact with other as well as other supply-led trends, such as the future of air transport and increasing demand from the middle classes in emerging economies outside of locations that have hitherto driven demand for global tourism. Forecasters face the challenge of trying to predict whether tourism and leisure demand will continue to grow in the unfettered manner seen prior to Covid-19, and, if it does, how this growth will be distributed by sector and geographically.

According to the UNWTO, in 2019 (the last year before Covid-19 distorted global travel patterns and demand), 1.46 billion people travelled globally – a 4% increase on the previous year. By continent, this aggregate figure comprised:

- *Europe*, which dominated arrivals, with 735 million (51% of the global total).
- *Asia-Pacific*, the second most important region, with 361 million arrivals (25% of the global total), including a large proportion of intra-regional travel (i.e. within Asia).
- *The Americas* (North and South America), with 219 million arrivals (15% of the global total), including significant travel across the US–Canada and US–Mexico land borders.
- *Africa*, with 73 million arrivals (5% of the global total). Although this figure has increased in recent years, it is still very small in relation to the continent's land mass, and arrivals tend to be concentrated in a handful of countries.
- *The Middle East*, with 61 million arrivals (4% of the global total). Like Africa, this region has experienced significant growth, with several states promoting tourism-related development to compensate for inevitable declines in oil revenues.

Yet, international travel is not the main driver of global tourism. As the WTTC pointed out, while 28.7% of all tourism activity is international, 71.3% is domestic. Similarly, the motivation to travel for tourism is broken down into 21.4% for business and 78.6% for leisure. There are debates over whether travel post-pandemic will recover to previous levels given the use of technology to facilitate business activity during lockdowns and periods of social distancing and travel restrictions.

The OECD (2018) referred to four specific megatrends impacting global travel demand:

- *Evolving visitor demand* as mature regions such as Europe and North America are supplanted by emerging regions such as Asia-Pacific, with its fast-growing middle class and new consumer groups, including millennials and Generation Z. It is within this context that ageing assumes a major role as an older population becomes a larger proportion in some mature tourism-generating regions, which in turn cascades to receiving regions as a subsequent ageing of visitor demand. Concerns are also being raised about the amount of disposable income available to future older consumers, who will probably have to work for more years prior to receiving less generous pensions (see Chapter 2). Other notable trends, such as longer life expectancy (and therefore more aged travellers), singles travelling in older age and multigenerational travel, all present both opportunities and challenges for an evolving tourism market, as Chapter 2 highlighted.

- *Sustainable tourism growth*, particularly in relation to the disruptive effects of climate change, which may pose greater risks for destinations and travellers so that hitherto popular destinations for ageing consumers become increasingly untenable. Similarly, governments may not allow continuance of the unmanaged resource consumption, poor eco-efficiency and unrestrained growth that have characterised global tourism since the 1950s.
- *Enabling technologies*, particularly new technologies, have been embraced by many generations, but older travellers in the future will be more tech-savvy and more comfortable with using technology to enhance and plan their experiences in ways that cannot even be imagined today. The challenge will relate to new technologies driven by artificial intelligence and robots, especially in ageing markets, where cognitive decline and other conditions may limit their adoption and use, regardless of the benefits.
- *Travel mobility* – that is, transporting people from home to destination, around the destination and back home – has always been the hallmark of efforts to make tourism an accessible experience. This connectivity role for tourism-related transport assumes far greater significance in older-age travel, where specific modes and access facilitate users' ability to break out of the home environment for leisure or tourism. The availability of key transport modes, such as air transport, may be more contested in the future as the reality of climate change highlights the problems of travelling for pleasure and leisure, and users are increasingly obliged to foot the bill for offsetting the environmental damage caused by transport for tourism. This may have a direct influence on the balance of domestic and international tourism, as demand typically adapts to the opportunities and constraints that governments create (e.g. by introducing visas and limiting access to travel and transport). In other words, demand may be mediated by state policies and the political environment. KPMG's (2019a) *Future Mobility* report outlined some other key changes that will impact the mobility market in terms of the expansion of electric vehicles, connected and autonomous vehicles and mobility as a service (MaaS), which will create a mobility ecosystem. The report (KPMG 2019a: 9) argued that 'Mobility-as-a-Service is an evolving concept of how consumers and businesses move away from vehicle ownership towards service-based transport. In this sense, MaaS includes multi-modal aggregation of transport modes as well as on-demand mobility.' In other words, future innovations will change the way personal vehicle usage is configured (i.e. the traditional buying, leasing and renting options, supplemented by car subscription services). Uber has already had a significant impact as an innovation offering an increased level of flexibility and accessibility for users through technology. In terms of automated vehicles, Harper *et al.* (2016: 1) suggested that these innovations could enhance the mobility of older people who are unable to drive:

> Many seniors (those over age 65) and people with medical conditions often face challenges travelling freely and independently and must rely on family, friends, government, or other providers to meet their basic

mobility needs. Automated vehicles represent a pathway that could increase the mobility, and hence the vehicle miles travelled (VMT), of the senior and disabled populations by decreasing human involvement during driving.

But, as one of our respondents (R9) indicated, there are substantial barriers to replacing one's own car with alternatives, such as taxis:

> Travel has been a really interesting one for us because it is a challenge for the older old group – the older old people whom we serve. Many of them will have got to the stage where they are giving up car driving and that is something which sometimes has to be encouraged because they are no longer safe driving … [but] there seems to be a great reticence to use taxi travel, even though many of our clients have sufficient money. They don't want to spend the money on travelling in a taxi, even though that might be a lot cheaper than [running and insuring] a car, as they did previously.

Scope and scale of the visitor economy: supply-side issues

Seeking to understand the scale and scope of the global visitor economy and its various components is a challenging task, as most conceptualisations tend to examine *either* tourism *or* leisure as independent sectors (see earlier chapters). We need to understand the 'supply' elements of what businesses and organisations provide in the whole visitor economy domain. Sessa (1983: 59) argued, 'tourism supply is the result of those productive activities that involve the provision of goods and services required to meet tourism demand and which are expressed in tourism consumption'. In effect, we are considering *both* the product (or experience) that is provided *and* how it is delivered to the consumer (now outlined as the supply chain). Perhaps one of the most useful overviews of this concept was provided by Tapper and Font (2004: 1), who described it as:

> all the goods and services that go into the delivery of tourism products to consumers. It includes all suppliers of goods and services whether or not they are directly contracted by tour operators or by their agents … or suppliers (including accommodation providers): Tourism supply chains involve many components … bars and restaurants, handicrafts, food production, waste disposal, and the infrastructure that supports tourism in the destination.

The emerging concept of Tourism Supply Chain Management (TSCM) has developed the notion of supply chain management within the wider visitor economy, with the aim of organising and optimising the efficient delivery of services and experiences. However, a contrasting approach exists within leisure research, in which the supply of leisure is broadly associated with all of the resources, settings and environments that are created or already exist to meet different leisure needs. As Kreutzwiser (1989: 21) pointed out, 'supply refers to the recreational resources, both natural and man-made, which provide opportunities for recreation. In other words, it is a complex concept influenced by numerous factors and subject to changing interpretations.' The resulting participation is derived from the relationship between leisure needs, wants and desires and resources as well as available opportunities. According to Mihalič (2003), the supply of leisure is rather unusual because:

- It is associated with a variety of goods and services, which may be connected to specific places or environments that determine the characteristics and forms of leisure that may be undertaken.
- It is made up of a complex amalgam of different suppliers (i.e. the commercial sector, public sector and voluntary bodies) who have a variety of objectives in relation to the resources, facilities or opportunities they provide.

One of the inherent problems of studying leisure is what Roberts (2004) described as its 'blurred edges' – that is, it is a subject with no clear boundaries. This difficulty is associated with what one considers leisure to be, or how we define it at an individual level, as well as its scope. What this means is that the breadth of leisure provision is so vast that drawing boundaries becomes almost arbitrary. In other words, the provision of leisure is in a constant state of flux and evolving due to the participation of multiple organisations and agencies. Roberts (2004) considered leisure as a series of interconnected industries (and organisations within each industry sector) that influence both home-based and non-home-based leisure. But it may be better to think of leisure supply and its provision in terms of the different spheres of influence that the interrelated sectors (i.e. commercial, not-for-profit and public) have on leisure. It is the interplay of these three different sectors that leads to the final supply of leisure in any given context. This is why combining tourism and leisure (as well as other sectors, such as events and hospitality) within the broader visitor economy sector has a great deal of merit when looking at a phenomenon such as ageing. Such an approach also means that arbitrary distinctions are not drawn between the various types of ageing users (i.e. leisure visitors versus tourists). In practice, all of these are visitor economy users, albeit with differing motivations and needs.

Haywood *et al.* (1989) produced a typology of six major leisure activities (recreation; hobbies, crafts and education; tourism and holidays; entertainment; commodities and shopping; gaming and gambling) on the basis of two overarching elements that affect supply:

- *The formal dimension* – that is, the characteristics or form of the activity, which may involve a level of active participation or simply passive consumption of leisure time. Ideally, during analysis of a particular leisure activity, researchers should collect information on participants' feelings towards the activity as they undertake it plus their reflections on it after the event.

• *The contextual dimension* – that is, the geographical location of the activity and the way in which and where it is provided and managed by the public or private sector.

The scale and significance of the visitor economy

Interestingly, there is an abundance of data on the scale and significance of tourism but rather less on those of leisure, even though leisure visits comprise the majority of visitor economy interactions in most countries around the world. The WTTC expects global tourism arrivals to reach 1.8 billion by 2030, driven by some powerful trends that are already shaping the wider visitor economy, including:

• An increasing interest in and demand for personal fulfilment, partially met through the introduction of new technology, such as smartphones;
• An increased ability on the part of guests to be able to create some degree of distance from work by moderating the influence of digital technology (e.g. smartphones) on their leisure experiences and removing their connection to the workplace;
• Increased demand for luxury experiences;
• Restructured work patterns, including the gig economy, which will create a more independent workforce and boost the sharing economy;
• Greater urbanisation of the population globally, leading to more empowered individuals; and
• More ethical consumers.

The WTTC predicts that these factors will help to generate a global experience economy worth US$8.2 trillion by 2028 (Fleetwood 2020).

Some of the key areas that are already affecting the sector are technology within the home (e.g. gaming with virtual-reality enhancements and online gambling) and out-of-home activities such as eating out (as well as home deliveries of restaurant food), cinemagoing and health and fitness (e.g., in 2016, Savills estimated that Europe's 80 million gym memberships were worth a total of €25 billion). Indeed, in total, the global wellness market, with its supply chain connected via tourist destinations or the day-to-day use of leisure facilities, was estimated to be worth US$4.2 trillion in 2016/17 (KPMG 2019b).

Yet, if growth in these areas is to be sustained as predicted, businesses and other stakeholders will have to meet the specific needs of ageing visitors at each of the touchpoints in the supply chain. However, as Page and Connell (2010) demonstrated, and as shown in Figure 4.3, such needs, which must be understood and met through the supply of suitable visitor economy experiences, cover a wide range. An extensive body of theoretical research has been developed within the field of leisure studies that seeks to explore how these needs can best be met. Many of these studies have looked at individuals' leisure behaviour and how

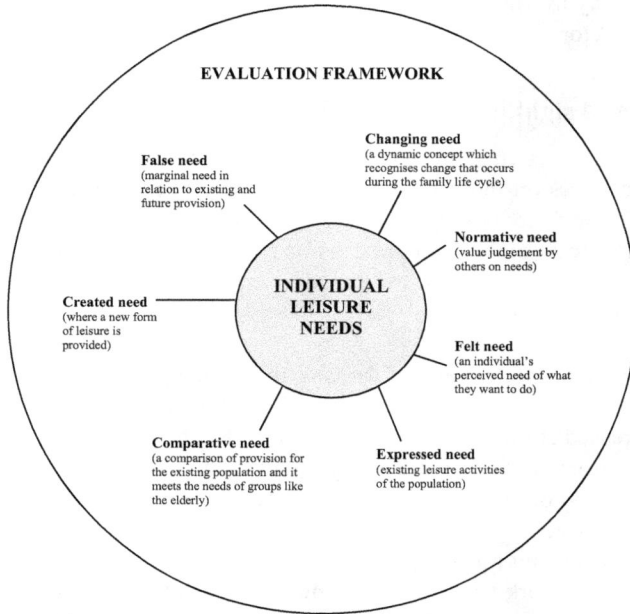

Figure 4.3 Model of leisure needs

Source: *Page, S; Connell, J; Leisure: An Introduction, 1st Edition,* © *2010. Reprinted by permission of Pearson Education Limited*

this changes through the life cycle, especially in older age (see Page and Connell 2010 for more detail); one of the factors conditioning leisure behaviour is disability.

Disability: definitions, debates and significance

Disability is an important feature of human society that has generated a great deal of debate, to such an extent that 'disability studies' is now an established academic field of study. Yet, as McPherson *et al.* (2020: 491) pointed out:

> disability is a contested term meaning different things to people and organisations, ranging in contexts (Erevelles 2011) and models: including the medical model (Areheart 2008), the social model (Oliver 2013), and the critical disability approach where there is a recognition that disability is socially constructed and that societal structures produce disability (Hughes and Paterson 2006).

Simpler overviews, such as that of Verbrugge (2016: 1124–5), describe the different facets of disability as they affect each individual, defining disability as:

'... difficulty doing tasks/roles on one's own due to health and lasting for some time' and encompassing '... physical, sensory, emotional or cognitive' health conditions. The widespread acceptance of the parameters of this type of definition has meant that most research has focused on daily living and household management. In contrast, the study of disability in relation to consumers is a relatively recent phenomenon (Burnett and Baker 2001), even though, as Blichfeldt and Nicolaisen (2011) demonstrated, forecasts suggest that there will be 100 million disabled people in the United States alone by 2030. Similarly, Cassia *et al.* (2020) reported that up to 20% of the UK's current population have a declared disability. The recognition of structural and design barriers and how these can be addressed, alongside an increasingly strong voice from voluntary bodies and communities, has promulgated these key issues to public-facing businesses and organisations.

Within the visitor economy context, Blichfeldt and Nicolaisen (2011: 79) pointed out that,

> although these consumers may have the freedom and financial means to travel, age-induced disabilities (either occurring to the consumer himself/ herself or to their spouse) may restrict leisure travel. As such, gaining knowledge on disabled tourists is crucial if the tourism industry wishes to develop products and services of value to many customers who have some kind of disability (or travel with companions with a disability).

In other words, disability should not be a barrier to participation in leisure and tourism, and there is a significant and growing market of ageing consumers with disabilities that will expect and seek better access and provision. Before we embark on our assessment of consumers living with disabilities, it is worth exploring in a little more detail what disability is, its significance as a construct and the debates over its conceptualisation, as illustrated by McPherson *et al.*'s and Verbrugge's contrasting definitions of the term.

Roulstone *et al.* (2012) provided an interesting contextualisation of the evolution of disability studies during the 1960s, associating it with broader debates on discrimination triggered by the civil rights movement in the United States and the gay rights movement in Western societies. Specifically, they suggested that disabled people were viewed as a marginalised, minority group that was suffering from disadvantage, and that disability should be considered as a social issue rather than just a medical one. Such an approach recognises that societal responses to disability can, in themselves, be discriminatory and cause even greater marginalisation. According to Darcy (2003), the term 'disabled' is indeed a social construction, created by other members of society and their attitudes, which exacerbates the marginalisation of people with impairments by casting them as fundamentally different from the 'able-bodied', a term widely used but subject to much criticism. Some disability researchers have attempted to counter this tendency by focusing on the provision of assistance and the removal of barriers with a view to emancipating the disabled and reducing their marginalisation. As discussed in Chapter 2, the ultimate aim of this approach is to overcome the social exclusion of each and every marginalised group so that their inclusion becomes normalised.

Blichfeldt and Nicolaisen (2011: 80) provided a useful summary of the recent history of disability studies:

> Most disability research is founded on either medical or social model perspectives ... and during the last 20 years, these two models have established themselves as the dominant discourses within disability research ... The medical model focuses on relatively stable and narrow notions of abnormality and deficiency in medical terms ... whereas the social model hinges on the premise that disability is socially constructed ... The medical model has been subject to much criticism because 'the power to define, control and treat disabled people was located within the medical and paramedical professions' ... The social model arose as a reaction against the medical model and was informed by disability activists and disabled people [themselves].

One highly contested aspect of the medical approach to disability is the WHO's (2001) International Classification of Functioning, Disability and Health (ICF). Table 4.1 outlines models of disability devised since the 1960s, and for each model identifies key components and definitions of the term. It also demonstrates that a number of these models were incorporated within the ICF, which has become 'a globally implemented statistical, clinical and scientific research tool – an international classification – as well as a conceptualisation of functioning and disability' (Bickenbach 2012: 55). Consequently, the WHO is guided by the ICF whenever it collects and collates data from UN member states. From an ageing perspective, the ICF views disability as something that has an impact on bodily functions and thus on individuals' activities and participation. It conceptualises the issue in a universal context where health conditions and disability are addressed in a standardised manner, and treats disability as a lived experience, giving due consideration to the sociopolitical and environmental factors that impact upon it (Figure 4.4).

As a classification system, the ICF is not without its critics, mostly on account of how it measures various elements and distinguishes between 'participation' and 'activity'. Yet, we should not forget that the social model helps us to understand ability in the context of disability, which in turn enables people to build their own competencies and confidence, and consequently take advantage of leisure opportunities. As Fullagar and Owler (1998: 449) argue:

> Leisure offers us an opportunity to experience something different or challenging, or simply pleasurable. However, it also provides the chance for a person to change their relation to themselves. That is, to develop a more positive narrative, which is to exercise power over oneself in a pleasurable rather than destructive or inhibiting way.

There are wide-ranging views on the relationship between ageing and disability. For instance, Gruenberg (1977) argued that the proportion of disabled people is increasing due to medical advances that have reduced mortality but left ever

Table 4.1 Models, components and conceptual meanings of 'disability'

Model of disability	Components of the model	Conceptual meaning of 'disability'
Nagi (1969, 1977, 1991)	• Pathology • Impairment • Functional limitation • Disability	'Pattern of behaviour that evolves in situations of long-term or continued impairments that are associated with functional limitations'
Social (UPIAS 1976; Oliver 1990, 1992, 1996)	• Impairment • Disability	'Limit or loss of opportunities to take part in community life because of physical and social barriers'
Verbrugge and Jette (1993)	• Pathology/disease • Impairment • Functioning limitation	'Disability is experiencing difficulty doing activities in any domain of life due to a
Institute of Medicine (Pope and Tarlov 1991; Brandt and Pope 1997; Field and Jette 2009)	• Disability • Pathology • Impairment • Functional limitation • Disability	health or physical problem' 'The expression of physical or mental limitation in a social context – the gap between a person's capabilities and the demands of the environment'
International Classification of Impairments, Disabilities and Handicaps (WHO 1992)	• Impairment • Disability • Handicap	'In the context of health experience, any restriction or lack (resulting from an impairment) of ability to perform an activity in the manner or within the range considered normal for a human being'
ICIDH-2 (WHO 1999)	• Body function and structure (impairment) • Activity (activity limitation) • Participation (participation restriction) • Contextual factors: environment and personal	'Disability is an umbrella term comprising impairments as problems in body function or structure as a significant deviation or loss, activity limitations as difficulties an individual may have in the performance of activities and participation restrictions as problems an individual may have in the manner or extent or involvement in life situations'
Quebec (DCP) (Fougeyrollas 1989, 1995; Fougeyrollas *et al.* 1998)	• Risk factors • Personal factors:– organic systems:integrity/impairment– capabilities: ability/disability	No conceptualisation of disability as such, but a model of the 'disability creation process':'An explanatory model of the causes and consequences of

(Continued)

Table 4.1 (Continued)

Model of disability	Components of the model	Conceptual meaning of 'disability'
	• Environmental factors:– facility/obstacle– life habits • Social participation/ handicap	disease, trauma and other disruptions to a person's integrity and development'
ICF (WHO 2001)	• Body function and structure (impairment) • Activity (activity limitation) • Participation (participation restriction) • Contextual factors: environment and personal	As in ICIDH-2:'Disability is a difficulty in functioning at the body, person, or societal levels, in one or more life domains, as experienced by an individual with a health condition in interaction with contextual factors' (Leonardi *et al.* 2006)

Source: *Bickenbach, J. (2012) 'The International Classification of Functioning, disability and health and its relationship to disability studies' in N. Watson, A. Roulstone and C. Thomas (eds) Routledge Handbook of Disability Studies, Abingdon, Routledge, 51–66.*

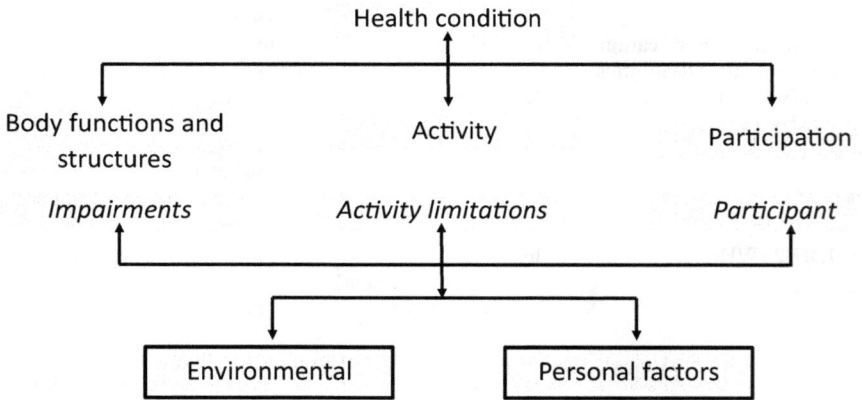

Figure 4.4 International Classification of Functioning, Disability and Health (ICF), devised by the World Health Organization (WHO 2001)

Source: Bickenbach (2012: 57)

more survivors with chronic conditions. In contrast, Fries (1980) suggested that these medical advances have 'compressed' disability by postponing it until later in life. However, this argument was criticised as too simplistic, given the range of complex medical conditions that may be associated with disability. Rather than focusing on the *number* of disabled people in a given population, Manton (1982) argued that new medical treatments are slowing the *progression* of disabilities. Once again, though, the data on this issue are extremely complex, so any attempt to present disability as a straightforward mild–severe continuum runs the risk of oversimplification. Lee *et al.* (2020) used longitudinal data sets to highlight considerable variations in the prevalence of disability around the world, pointing out, 'disability is a binary variable, indicating any difficulty in at least one of the activities of daily living' (Lee *et al.* 2020: 164). Yet the relationship between ageing and disability is extremely complex. For example, when analysing data from the United States and England, Lee *et al.* found high rates of disability in the over-60s but rather lower rates in the over-80s. In contrast, there seemed to be a direct and consistent correlation between ageing and disability in South Korea, Belgium, the Czech Republic (Czechia) and Mexico.

Amid all this complex and sometimes contradictory data, one fact remains clear: public health interventions, preventive medicine and new treatments are undoubtedly increasing life expectancy around the world (Newman *et al.* 2020). However, as Gruenberg (1977) predicted more than four decades ago, this has contributed to a significant rise in the number of people living with chronic diseases and disabilities. The international community has adopted a number of measures with a view to increasing longevity, including promoting healthier diets and more exercise, and eliminating smoking. Yet, those who follow this advice and consequently live longer are more likely to experience delirium, sarcopenia (loss of skeletal mass and function), incontinence, deterioration of physical and cognitive function and falls (which are often associated with frailty). For instance, in their analysis of ageing and disability in the United States, Molton and Ordway (2019) found that the average age of people living with some form of disability is increasing year by year. Meanwhile, La Plante (2014) estimated that between 12 million and 14 million US adults were ageing with a disability. As a result, epidemiologists are increasingly analysing big data, clinical trials, deep phenotyping (a new form of precision medicine, such as diagnosing dementia on the basis of neuroimaging data) and cohort studies in the hope of understanding how geriatric-specific medical conditions are impacting mobility, independence, social engagement and physical activity in older age groups.

Clarke *et al.*'s (2008) study of mobility, disability and the urban environment established that people with mild impairments encountered few significant obstacles. In contrast, those with profound neuromuscular and movement-related conditions were severely impacted by cracked pavements and broken kerbs. In addition, porches and steps increased their risk of falling, which is one reason why step-free access is a hallmark of UD (see Plates 4.1, 4.2, 4.3 and 4.4). Clearly, then, urban authorities should heed Verbrugge and Jette's (1993: 3) message that 'disability can be diminished swiftly and markedly if the physical and mental demands of a given task are reduced'.

However, this brings us back to the debate over what constitutes a disability. While some countries have reported declines in the number of people who are living with 'severe disabilities', there have been parallel rises in 'chronic conditions', such as lung disease, diabetes and cardiovascular disease, all of which increase the pressure on an already overstretched long-term care system. Obesity (Wild *et al.* 2004; Zamboni *et al.* 2005; Alley and Chang 2007), high blood pressure (hypertension) and severe hearing impairment are other ongoing challenges.

Plate 4.1 Certain aspects of the outdoor environment, including steps, can pose significant challenges to ageing visitors; this visitor attraction makes it clear that the
site cannot be made more accessible and that it is impossible to accommodate
wheelchairs

Consequently, as Beard and Bloom (2015) argue, there is an increasingly urgent
need for a comprehensive public health response to our ageing population.

Disability and the visitor economy

Disability and tourism

According to Blichfeldt and Nicolaisen (2011: 81), disability encompasses a
broad spectrum of issues, given that 'one can differentiate between mental and
physical disabilities; furthermore, physical disabilities may be divided into hearing, vision and mobility disabilities'. Irrespective of the precise nature of the disability, however, early studies found that very few disabled people were prepared
to engage in tourism, even though engagement became easier for those who were
willing to persevere, as they learned how to negotiate barriers and obstacles in
the supply chain. Disabled people are still much more likely to engage in leisure, rather than tourism, as this enables them to return to the security of their
home environment each day. Nevertheless, Darcy *et al.* (2020: 140) identified a
surge of recent interest in accessible tourism and the development of destination

Plate 4.2 Where there are minor barriers, such as using a standard wheelchair or taking those with limited mobility on to a beach, innovations such as beach-suited wheelchairs can be used

experiences that are able to accommodate each and every type of visitor equally well.

Up to now, research has tended to focus on specific aspects of disability and tourism, such as accommodation (Adam 2019; Darcy 2010; Piramanayagam *et al.* 2019; Tutuncu and Lieberman 2016), the sharing economy (Boxall *et al.* 2018; Randle and Dolnicar 2019), transportation and travel needs (Darcy 2012b; Ray and Ryder 2003; Shaw and Coles 2004), visitor attractions (Cloquet *et al.* 2018; Mesquita and Carneiro 2016; Poria *et al.* 2009), sightseeing (Dann and Dann 2012), societal attitudes to disability (Daruwalla and Darcy 2005), specific disabilities (Devile and Kastenholz 2018; McIntosh 2020; Moura *et al.* 2018; Pagán 2020; Richards *et al.* 2010; Sedgley *et al.* 2017; Small *et al.* 2012; Zajadacz 2014), information provision (Eichhorn *et al.* 2008; Shi 2006), country studies (Agovino *et al.* 2017; Figueiredo *et al.* 2012; Gregoric *et al.* 2019; Ozturk *et al.* 2008), parents travelling with disabled children (Freund *et al.* 2019; Kim and Lehto 2013; Nyman *et al.* 2018; Tecau *et al.* 2019), the language used in enabling disabled tourism (Gillovic *et al.* 2018), disability tourism's relationship with social inclusion (Kastenholz *et al.* 2015), mobility issues (Kim and Lehto 2012; Var *et al.* 2011; Yau *et al.* 2004), destination design (Lam *et al.* 2020), theoretical

Plate 4.3 In an ideal world, attractions would have flat, easy access in order to accommodate most users

analyses (Lee *et al.* 2012), pricing and cost-effectiveness (Lyu 2017; Morad 2007; Pagán 2012) and contributions to quality of life (Pagán 2015).

The wide range of research studies indicates the importance and prevalence of disability as a key area of interest, policy and practice. Darcy *et al.* (2020) estimated that between 10 and 20% of the world's population now have some sort of disability, and suggested that disabled travellers are worth some 50 billion euros to Europe's visitor economy, and thus represent a highly significant sector of the total tourism market. Nevertheless, Cassia *et al.* (2020) identified a range of barriers that people with disabilities still face when trying to access a destination:

- Knowledge of and information about the destination;
- The built environment, specifically issues with walkways and pavements;
- A lack of political commitment to addressing these obstacles in the built environment;
- Societal disregard for and even antipathy towards people with disabilities, coupled with a fundamental lack of understanding of disabled people's capacity to lead fulfilling lives; and
- Entrepreneurial myopia as businesses fail to appreciate the potential of the disabled tourism market (see Connell *et al.* 2017; Connell and Page 2019a).

Plate 4.4 Newer attractions developed with accessibility in mind, such as the Eden Project (which works closely with the Sensory Trust and has won awards for its commitment to accessibility), have specially designed infrastructure to make the site accessible to all as well as pre-bookable standard and powered wheelchairs and all-terrain mobility scooters

Disability and leisure

Research into the leisure industry's relationship with disability has a longer history than the corresponding tourism studies literature. In one important study, McLachlin and Claflin (2004) stressed that the visitor economy must take a range of different disabilities into consideration, including:

- *Developmental disabilities*, including chronic conditions relating to a particular mental or physical impairment (or a combination of the two);
- *Physical disabilities*, which may be associated with physical degeneration due to chronic health conditions, congenital malformations, skeleto-muscular disorders and/or neurological impairment;
- *Hearing impairment*, the world's most common disability, with around 22–28 million hearing-impaired people in the United States alone (plus millions of others who do not wish to be identified as hearing impaired and so are not included in official statistics);

- *Visual impairment,* including partial sight, limited vision and complete blindness; and
- *Emotional disabilities,* including serious psychological conditions, such as schizophrenia, and personality disorders that make it difficult for sufferers to form relationships and enjoy leisure.

As Page and Connell (2010) argued, all disabled people must be able to choose when, where and how they participate in leisure. For example, some may prefer to spend most of their leisure time with friends and other people with similar conditions; and sporting events such as the Paralympics obviously have strict rules regarding participation to ensure a level playing field for all competitors. On the other hand, Aitchison (2003: 956) defined leisure as:

- Playing an important role in increasing self-esteem, confidence and psychological well-being;
- Enhancing physical health and fitness;
- Reducing the risk of illness;
- Contributing towards positive social interaction and relationships.

Therefore, mingling with able-bodied people in leisure activities can help disabled people to overcome feelings of social disconnection, loneliness, sedentary lifestyles and many of the issues associated with deprivation and a low quality of life (Bullock and Mahon 2001). However, to achieve this, we need to understand and address the constraints that disabled people face when they attempt to participate in leisure activities (Burns and Graefe 2007). Some progress has already been made in this area. For example, Scotland's Forest Enterprise has provided sensory trails for the visually impaired, while the British Sports Association for the Disabled, the UK Sports Association for People with Learning Disability and Riding for the Disabled have all sought to increase participation in their respective areas.

According to McPherson *et al.* (2020: 494), Critical Disability Theory (CDT) has an important role to play in understanding how a more emancipatory approach can be developed to facilitate greater inclusivity:

CDT adheres to the notion that social and environmental barriers, such as inaccessible buildings and transport, discriminatory attitudes, and negative cultural stereotypes, are 'disabling' people with impairments ... Persons with disabilities may experience functional limitations that non-disabled persons do not experience, but the biggest challenge comes from mainstream society's unwillingness to adapt, transform, and even abandon its 'normal' way of doing things.

This theory, which is equally applicable across the entire visitor economy, highlights four domains where urgent action is needed:

- Recognising the contribution of leisure and changing societal attitudes to enable the non-disabled population to appreciate the abilities of disabled people and the contribution that participation in the broader visitor economy can make to their quality of life. This has an added dimension for older people with disabilities, as it can help to break down exclusionary barriers and thereby reduce isolation, loneliness and boredom while enhancing social engagement.

*   Ensuring that policies and strategies are supported with the necessary resources to enhance access and reduce discrimination.
*   Providing accessible information in suitable formats, such as Braille, and using simple language.
*   Ensuring unrestricted physical access to leisure spaces and other places by adhering to the principles of UD.

McPherson *et al.* (2020) also highlighted the range of obstacles that major sporting events will need to remove in order to comply with the UN Convention on the Rights of Persons with Disabilities (UN 2018). Similarly, Darcy (2012a) outlined the measures that those organising events must take to make them fully accessible for disabled visitors, while other studies have focused on disabled volunteers at the Olympics and Paralympics (Dickson *et al.* 2018) and participation rates of people with cognitive impairments (e.g. Tint *et al.* 2017). However, very few studies have explored disability and ageing in the visitor economy.

One of the few studies to address this issue from a broader social science (as opposed to medical) perspective is that of Engeland *et al.* (2018), who note that 'The experience of transitioning to retirement is highly individual in all populations and can depend on whether retirement is voluntary or involuntary and how well prepared individuals are for the transition' (Engeland *et al.* 2018: 73). Twenty years earlier, in their study of people with intellectual disabilities, Rogers *et al.* (1998: 122; emphasis added) observed that 'The most pronounced theme that emerged from the data was lack of self-determination in leisure. Participants had few opportunities to freely choose leisure in any aspect of their lives. In many cases, opportunities for self-determined leisure were further constricted by age-related changes in the participants' lives.'

There is much to learn from studies that adopt a life-course approach to disability and ageing (e.g. Jeppsson Grassman and Whitaker 2013), which view people with disabilities as individuals, as opposed to a homogeneous group. It is also judicious to consider the environment with which disabled people interact and how specific impairments can be overcome through enabling tools and actions. As Kelley-Moore (2010: 96) lamented:

The discourse about ageing and disability is dominated by the perspective that functional decline is a normative part of the human ageing process leading inevitably and irreversibly to disability. Such an understanding is so pervasive in social, medical, and policy domains that it is considered to be axiomatic. Scholars have argued that such a perspective exemplifies ageism because it treats chronological age as a necessary and sometimes even causal factor in the disablement process, creating a pessimistic and flawed view of ageing.

At this point, it is worth summarising the critical themes that have emerged from this overview of disability and the visitor economy:

- The relationship between ageing and disability is complex, so we should not fall into the trap of considering people with disabilities as a homogeneous group whose participation in leisure activities will be increased through the introduction of a simple checklist or information booklet.
- People with disabilities often have complex care and health needs throughout the ageing process.
- Visitor economy businesses should be cognisant of increasing frailty and other impacts of the ageing process, such as disability, and each person should be treated as an individual with individual needs.
- We need to look past optimistic notions of successful ageing and recognise that an increasing number of people are living with chronic conditions. Nevertheless, we should also be aware that most people strive to maintain an active, independent life and ageing does not necessarily equate to disability or poor health.
- The ageing process can exacerbate certain disabilities over time but, once again, it must be remembered that each case is different.
- Society's perceptions and assumptions with regard to disability, coupled with prejudice about ageing, can be just as problematic as physical barriers.
- Many disabilities are 'hidden', with no visual or physical cues.

For the visitor economy, the main challenge is to increase understanding of the multifaceted nature of disability. As this chapter has stressed, each individual disabled visitor's needs are unique, but it is still useful to explore the strategies that businesses may adopt to improve access for particular groups. Hence, the remainder of this chapter discusses some of the most beneficial approaches to several specific disabilities, starting with visual impairment.

Sensory issues

Visual impairment

According to the WHO (2019: 3), 'Vision is the most dominant of the five senses and plays a crucial role in every facet of our lives', especially later in life, when good vision helps people to maintain 'social contact, independence and mental health'. The same report suggested that 2.2 billion people are currently suffering from some form of visual impairment or blindness, and explained that 'ageing is the primary risk factor for many eye conditions. The prevalence of presbyopia [difficulty seeing objects close up], cataract, glaucoma and age-related macular degeneration increase sharply with age' (WHO 2019: 8). Indeed, statistics confirm that the main causes of blindness are: cataracts, uncorrected refractive errors and age-related macular degeneration (Bourne *et al.* 2013). Hence, Klaver *et al.* (1998: 653) concluded: 'The hierarchy of causes of blindness and visual impairment is highly determined by age. As yet, little can be done to reduce the exponential increase of blindness.' Nevertheless, many studies have shown that the onset of blindness need not exclude, nor even significantly limit, engagement with the visitor economy.

For the ageing population, holidays provide 'opportunities for social interaction and relaxation; experiencing different climates, cultures and countries; and generally a change from routine environments ... Of course, when individuals are deprived of one sense, they rely more heavily on their remaining ones' (Richards *et al.* 2010: 110). Yet, as Richards *et al.* (2010) point out, people with visual impairments tend to stay at home and miss out on the multi-sensory enjoyment associated with tourism because of the anxiety and stress of travelling in a sighted world that does not readily accommodate their needs, notwithstanding the accessible tourism guidelines and regulations that are now in force, particularly with regard to the aviation industry (Abeyratne 1995).

Enabling organisations such as the UK's Traveleyes, which plans holidays in which blind people are paired with sighted people, are working hard to overcome these barriers. Nevertheless, as one blind traveller pointed out:

What I find frustrating is that those systems aren't always in place on the cheapest form of transport – take buses for example. We definitely need 'talking buses' in the UK – they announce which stop you're at, just like trains or trams or the Tube. It's not rocket science, but it really helps. If I take the bus in the UK I have to ask the driver to tell me. If they forget, then you have to get off somewhere completely unfamiliar and rely on someone else helping you. It's very frustrating.

(Quoted in Plush 2017)

Richards *et al.* (2010: 1112–13) emphasised that people with impairments do not comprise a homogeneous group; indeed, they have just as many differences and idiosyncrasies as the general population. Other studies of visually impaired tourists (e.g. Poria *et al.* 2011) have highlighted the importance of listening to each consumer's voice and suggested ways in which organisations can help such travellers to retain their dignity. Meanwhile, Hersh's (2016) analysis of deaf/blind travellers provided important advice for tourism organisations and travellers alike. Similar advice is contained in the Royal National Institute for Blind People's factsheet on welcoming blind customers (www.rnib.org.uk/sites/default/files/Welcoming-your-blind-and-partially-sighted-customers-tourism-factsheet.doc). Similarly, Packer *et al.* (2008) discussed how best to advertise an attraction to visually impaired potential visitors; procedures to follow at the point of entry, on site and at departure; hospitality issues; and site-specific measures, such as the installation of tactile surfaces. There are many examples of good practice, such as: the Typhlological Museum in Madrid, which was built with no barriers, following consultations with the National Organisation of the Spanish Blind (ONCE); the Science Museum in London, which is equipped with Braille interpretation; Cadbury's World in Birmingham, where blind visitors can touch props and listen to audio tours; and Manchester's Chill Factor, home to Disability Snowsport UK, which hosts adaptive activities for people with various disabilities, including sensory issues.

One of our respondents (R3), who works for a charity that supports visually impaired people, highlighted technology's recent impact on their lived experiences:

> We've got about twelve thousand people on our Facebook Connect community, all blind or partially sighted people, and that is growing by about three thousand a year. And ... the biggest demographic is not younger people. It is older people who are really tech savvy, you know. That's the difference. I think people who may have been at work with a mobile phone at 60 or 65, they don't suddenly go, 'Oh, now I'm retired, I'll give up my technology.' They've got very different expectations to, say, a group who are ten years older than them. People who are now in their mid-to-late 70s know a huge amount about technology ... There's also loads of tech-savvy late 70s, 80s, 90s ... Actually, it's just ingrained in the DNA of people who are slightly older, and that is changing things so much for information, for independence, for talking to other people. It's really interesting: technology really is the key for that confidence and that independence and tackling isolation.

The same respondent discussed hotels that have been specially adapted for people with sight loss:

> We used to have three hotels across the country which were specifically for older blind or partially sighted people. But there was a shift away from people wanting that; some people felt it was a ghetto, you know. They were all there together and, actually, they wanted [to be in] a normal, mainstream hotel with everybody else, Joe public. But we do still organise – and support volunteers who organise – events, day trips, social groups. We've got groups who do astronomy, even for people with sight loss, we've got LGBTQ groups, we've got ... language groups – Spanish language groups, Polish language groups, Welsh language groups ... Some of them, when they can do face-to-face ... they meet up, they go to places.

Meanwhile, in the developing world, Adam *et al.* (2017: 317) found that 3% of Ghana's population have some sort of disability, including 'visual, hearing, speech, physical, intellectual, emotional, and multiple disabilities. Disabled people in Ghana generally experience social exclusion, stigma, and marginalisation.' Within this study, 61% of the sample had a visual impairment, and 25% of all respondents were over the age of 60. Eleven important leisure activities emerged during the course of the research: conversation (chatting), listening to music, watching television, listening to the radio, sleeping, visiting friends and relatives, meditation, reading, playing ludo, playing oware (a local board game) and playing draughts (see Table 4.2). Many visually impaired respondents indicated that

Table 4.2 Leisure pursuits among Ghana's disabled population

Leisure activity	N	Per cent	Rank	Average time (hours)
Chatting	278	16.8	1	2
Listening to music	251	15.2	2	2.8
Watching television	231	14.0	3	2.6
Listening to the radio	187	11.3	4	2.4
Sleeping	164	9.9	5	2.3
Visiting friends/relatives	151	9.2	6	1.9
Meditation	148	8.9	7	2
Reading	77	4.7	8	2.4
Ludo	61	3.7	9	2.1
Oware	53	3.2	10	2.3
Draughts	51	3.1	11	2.5
Overall	1652+	100		2.3

Note: Multiple-choice answers.

Source: Adam *et al.* (2017: 322)

they were motivated to participate in one or other of these activities by a desire to master a specific skill.

Hearing loss

Age-related hearing loss (*presbycusis*) is characterised by difficulties in understanding speech in noisy environments. Three tools – hearing aids, cochlear implants and lip-reading or sign language – may help with this. Ho and Peng (2017: 449) explain that 'hearing impairment is defined as the inability to hear sounds at a normal level in decibels. In other words, people with hearing impairment have difficulties in receiving voice messages, and they encounter communication difficulties throughout their lives.' However, we need to consider a much broader range of issues when addressing hearing impairment in the visitor economy, as Zajadacz's (2012: 434) excellent overview illustrates:

> The term deaf is defined both in a medical and social context. From a medical point of view, a person is audiologically deaf or has a profound hearing loss if he/she has a pure tone average equal to, or over, 95 dB HL … People with a hearing loss below 95 dB HL are often known as hard-of-hearing … From a cultural perspective, a person is considered culturally Deaf if he or she views himself/herself as belonging to a cultural minority, that is, the Deaf community … and uses sign language as the main mode of communication … The degree of hearing loss is not of importance. An uppercase letter is sometimes used to separate culturally Deaf from other deaf people … A common distinction made when writing about deafness is between 'deaf' and 'Deaf', the former refers to deafness solely as an audiological condition, the latter to deafness as a cultural condition.

In other words, as with other disabilities, the term 'deafness' encompasses a broad spectrum of people, and societal attitudes have an important role to play in how those with any degree of hearing loss are viewed and treated. From a medical perspective, Blume (2012: 350) pointed out that 'in many cases of deafness, the problem lies in the inner ear; or cochlea. It is in this snail-shaped organ that sound waves are converted into electrical stimuli that are carried to the brain.' Such problems increase with age. For instance, in the United States, a third of people between the ages of 65 and 74, and half of those aged over 75, have suffered at least some loss of hearing. Nevertheless, as Gething (2000) pointed out more than two decades ago, policymakers have routinely neglected this significant sector of society, and businesses have frequently done the same. Gething also stressed that those who have been deaf throughout their lives tend to develop effective coping methods, whereas those who develop hearing loss in older age often find it much more problematic. Yet, the existing literature on hearing loss and the visitor economy has had very little to say on the subject of older tourists (e.g. Var *et al.* 2011). In contrast, there has been ample research into the experiences of younger deaf travellers. For example, Ho and Peng (2017: 449) identified the value of backpacking for members of the deaf community:

> Five push themes (constraints of group tours, self-challenge, independence, different experience, and invitation by hearing-impaired friends) and two pull themes (enjoy local culture and lifestyle and the 'I have been there' feeling) were identified. Furthermore, the participants' desire to travel as backpackers is based on their previous negative experiences in group tours. Moreover, they prefer backpacking with hearing-impaired partners than with normal-hearing partners.

Atherton (2012) identified similar experiences in post-war deaf clubs and communal leisure activities, when hearing-impaired people's participation was influenced by those around them.

Figueiredo *et al.* (2012) surveyed 200 people with hearing, motor, visual and intellectual impairments and concluded that overcoming physical and attitudinal barriers and constraints required a range of activities, services and equipment. Conversely, though, Peng *et al.* (2013: 531) examined

> [the] travel behaviors of hearing-impaired people participating in group tours and found that hearing-impaired tourists with communication difficulties trust other hearing-impaired travelers more than they trust those without a hearing impairment, and they prefer to obtain travel information from the Internet or friends rather than from travel agencies.

A number of country-specific studies dominate the literature on deafness and the visitor economy, with some countries examining broad domestic travel issues for the hearing impaired (e.g. the Canadian Hearing Society 1989). One of the key issues for the visitor economy, as we can deduce from studies such as Zajadacz (2012), is that while the Internet is a useful travel-planning tool, deaf people's main form of communication remains sign language. This implies not only that the visitor economy should prioritise the learning of sign language among employees, but also that alternatives, such as the graphical representation of information, should be developed to make travel more accessible for hearing-impaired people. A number of tourism operators, such as Responsible Travel, have already embraced these notions, as its mission statement testifies:

> Hearing impairment may be a 'hidden disability' but it's one about which increasing numbers of tourism operators and destinations are becoming significantly more aware. At Responsible Travel we believe that having difficulty hearing, or being unable to hear anything at all, shouldn't mean you should be prevented from travelling the world, and it also shouldn't mean that your experiences while doing so should be any less rewarding than those that can hear well. To that end, we work with our holiday companies to learn about the different ways they can accommodate guests that are deaf or have limited hearing, and we encourage them to keep doing more wherever possible … As long as suppliers have sufficient advanced notice that you are deaf or hard of hearing, many activities can be successfully adapted so that you can take full part.
>
> (www.responsibletravel.com/holidays/accessible/travel-guide/
> deaf-and-hard-of-hearing)

In New Zealand, the country's National Foundation for the Deaf sponsored a survey of 365 holidaying deaf people, with 52% of the domestic travellers and 34% of the international visitors in the sample aged over 60 (AUT 2011). Unfortunately, there was no bivariate analysis of the data with age as a controlling variable, so we cannot deduce much from this survey with regard to the specific challenges that elderly deaf people encounter while on holiday.

Eberts (2017) stresses that all deaf travellers, regardless of age, should prepare well in advance, inform tour guides of the extent of their impairment and make full use of technology:

> Having your hearing aids on the fritz can be troubling at anytime, but when you are far away from home and your audiologist – in another country, for example – it can feel like a disaster. Set a back-up plan before you go and test it out so you can easily implement it if needed. Examples include using a pocket-talker, an FM system, or connecting a high-quality headset to an app like EarMachine on your smartphone. If you have spare hearing aids, bring those too.
>
> (Eberts 2017: 42)

In the UK, the British Deaf Association's *Guide for Tourist Associations* is designed to help visitor attractions accommodate deaf people (https://bda.org.uk/guide-for-tourist-organisations/), as is VisitEngland's *Listen Up: Tips and Advice to Help you Welcome Customers with Hearing Loss* (www.visitbritain.org/sites/default/files/vb-corporate/Documents-Library/documents/England-documents/listen_up.pdf). The latter was produced in conjunction with the charity Action on Hearing Loss.

Learning impairments and autism

According to Patterson and Pegg (2009b: 389), 'intellectual disability has been characterised by evident limitations in intellectual functioning and adaptive conduct, the latter expressed as it relates to conceptual, social and practical adaptive skills'. Nevertheless, it often manifests as a 'hidden' disability as there may be few, if any, visible signs of learning impairment. Unfortunately, there has been very little research into this group's engagement with leisure and tourism.

In contrast, autism has attracted a great deal of interest, especially in recent years (e.g. Thurm and Swedo 2012), although most of these studies have focused on families with autistic children (e.g. Walton 2019), with very little thought given to the problems faced by ageing autistic adults, as Sedgley *et al.* (2017) point out (see also Hamed 2013). This is hardly surprising for two reasons: first, autism is probably the world's fastest-growing development disability, with Baron-Cohen *et al.* (2011) reporting that it now affects 1.31% of all eight-year-olds in the United States; and, second, people with autism's life expectancy is roughly 30 years shorter than that of the general population, so, sadly, many do not reach old age (see Heslop *et al.* 2013). Stacey *et al.*'s (2019) analysis of autistic adults in Australia concluded that leisure participation is a key factor in improving their well-being, but also that they tend to be less satisfied with their leisure time than the general population. Wright (2016) explored the specific challenges faced by autistic adults in greater depth, but had little to say on the subject of their engagement with leisure and tourism.

As with every other disability that we discuss in this chapter, the term 'autism' encompasses a broad spectrum of conditions; in the case of autism, mostly involving repetitive behaviour and difficulties with social interaction, with symptoms ranging from mild to severe. Povey *et al.* (2011: 230) provide a useful summary:

> Autism is a lifelong condition, which is now thought to affect one in 100 people … While it was first recognised as a distinct syndrome in the 1940s by Leo Kanner, it was not until 1975 that the concept of autism as a spectrum disorder was described by Lorna Wing and her colleague Judith Gould. Their work recognised that autism affects each person differently, and that though all people with autism have defined characteristics necessary for a diagnosis, they will vary in severity and complexity. Those key characteristics necessary for diagnosis are known as the 'triad of impairments' relating to difficulties with communication, social interaction and social imagination. The core problems in autism relate to the development of the social instinct (the ability to relate to others) in childhood. Consequently as individuals with autism develop, they usually find huge problems in social understanding and in their relationships with others. For example, they may find no pleasure in sharing company with others, or may wish to interact but struggle in their understanding of the usual social cues and behaviour. This also affects their ability to communicate with others and to make sense of the world around them, particularly in the context of behaviours, motivations and the inner

thoughts of other people. Their thinking is likely to be inflexible, and they may have a restricted and repetitive repertoire of activities … Some will need high levels of support throughout their lives, have no language, and very severe challenging behaviours; others will be able to attain academic qualifications and achieve great things in their chosen career, though they may continue to struggle with social relationships and the complex social interactions happening around them.

Identifying, understanding and accommodating people with autism is a major challenge for the visitor economy, not least because most of the interactions in the tourism and leisure sectors take place outside the safety, security and familiarity of the family home (see Sedgley *et al.* 2017; Gray 1994). Specific issues include extreme unease about travelling by air (see Graham *et al.* 2019), an inability or unwillingness to socialise, especially with strangers, a focus on hobbies and interests to the point of complete immersion and heightened anxiety in new settings. As one might expect, Stacey *et al.* (2019) found that people with autism tend to favour solitary activities.

Despite these difficulties, there has been some progress with regard to improving autistic visitors' access to tourism and leisure facilities. For instance, the National Autistic Society and VisitEngland (2018) have published a guide that provides five tips for dealing with autistic visitors:

1. Be patient and give people space during a meltdown.
2. Notify people of any changes to services.
3. Help to alleviate social anxiety.
4. Give people plenty of processing time.
5. Take steps to avoid sensory overload.

Other organisations, such as Autism Travel and Tourism for Autism, offer autism-friendly holidays and advice to other businesses in the visitor economy; Warner Studios has an 'autism guide'; and Newcastle United provides stadium tours for people with autism. Such initiatives are important as they shift the emphasis from how people with autism comport themselves to how their behavioural needs can be accommodated to provide them with a rewarding and memorable experience. As Tourism for Autism explains on its website (www.tourismforautism.com):

> People with autism need structure and predictability. When going on a vacation or day trip, this structure often completely disappears. This can result in stress and even in avoiding such activities. Tourism for Autism tries to bring about change by offering support to tourist organizations that come into contact with this target group during their holiday or trip.

Similarly, Tourism for All (www.tourismforall.co.uk/autism-friendly) has gone to great lengths to identify accommodation options and locations that are suitable destinations for people with autism.

Summary

This chapter has outlined the importance of conceptualising the way in which we understand the experiences of ageing visitors and their interactions with the visitor economy, with particular reference to the barriers they face due to medical conditions associated with the ageing process (e.g. disabilities and sensory issues). Such conditions pose some serious challenges for leisure and tourism businesses in terms of making destinations more accessible to ageing visitors. It is clear that a multidisciplinary (and ideally an interdisciplinary) approach is essential if the visitor economy is ever to overcome these challenges. The key message that has emerged in the course of this chapter is that people should always be treated as individuals with nuanced life experiences, not as statistics, nor as homogeneous groups, regardless of the medical conditions or impairments they may or may not share with others. As we have seen, professional organisations that specialise in understanding and overcoming barriers to full access and engagement (such as deaf associations) can provide invaluable help to leisure and tourism businesses by explaining how they might adapt to accommodate each and every visitor (often without the need for costly investment). This theme is explored further in Chapter 5, which focuses more explicitly on age-related conditions and the associated challenges that these create.

5 Accommodating visitors with specific needs

Perspectives on mental health, physical and degenerative conditions

Introduction

In the previous chapter, we examined the concept of disability and a range of conditions that challenge the ability of people to engage with the visitor economy. This chapter continues with that focus, further exploring the interactions between ageing, physical, cognitive and mental health conditions and the visitor economy, including the types of advice needed, such as travel medicine. This chapter will examine the emotive mental health issue of loneliness and ageing, a feature that is periodically aired in the popular media especially around holiday periods such as Christmas, when family separation may be most acute. A study by Age Cymru (2020: 9) during the Covid-19 lockdown found that 'one third of older people have been lonely … 55% of older people living by themselves were lonely during lockdown … [and] … half of older people living with a disability were lonely during lockdown' in Wales. In April 2020 the government provided £5 million of funding for national loneliness organisations to address the specific effects of loneliness that were being compounded by the national lockdown. The wider experiences of loneliness (accentuated by Covid-19) are a reflection of a changing society that, as this chapter highlights, has seen the dilution of the notion of extended family in many developed societies since 1945. This section is followed by a broader discussion of the issues of loneliness, mental health and depression, and the role that the visitor economy can play in addressing these prevalent societal conditions. We challenge the notion that taking a holiday is a simple solution to depression and other mental health issues for older people, and explore why its long-term value for that purpose is contested among researchers. This is followed by a discussion of physical health conditions that influence people's engagement with the visitor economy; we conclude with a discussion of degenerative conditions, with a specific focus on dementia.

Current issues and trends in visitor needs

In Europe, Deuschl *et al.* (2020) observed the trend towards a growth in neurodegenerative diseases, but a decrease in the rates of stroke and infections among the

DOI: 10.4324/9781003039358-5

ageing population. The growth in neurological disorders in Europe was higher in men than in women, clustering among individuals aged 80–84 years, although this varied by country and region. One element of the growth in neurodegenerative disease is the rise in dementia, which has replaced heart disease in the UK as the leading cause of death for an ageing population. Death is a reality as an end stage of the life course, and although it has a major sociological impact on families and their social networks (Cox and Thompson 2021), it is not a subject that attracts a great deal of discussion. This is demonstrated in the following quotation from one of our respondents (R1):

> Sometimes we organise holidays and we do it through like coach operators and people like this and again, it has to be very age appropriate … You have to think about things like relative insurance, you know, your procedures if someone, I mean, as has happened to us, dies on the tour … We had one tour where a lady died and as you can imagine, you know, it's a four-day holiday, Day 2 someone is found dead in the bed … for everyone else how do you recover the holiday from that. These things happen for any age group but it's not normally a risk as it is for the older adult, you know.

As the quotation indicates, death is a key element of human life, but it can also pose many logistical issues for the visitor economy. Although death is a taboo subject for many, it is a phenomenon that the visitor economy is not exempt from and needs to be prepared for, as Reid's (2017) analysis of the global epidemiology of tourist fatalities illustrates. Reid found that between 2013 and 2015 there were over 3000 tourist deaths globally, though not all will be recorded as such because the data sources used to corroborate the scale of tourist deaths may not cover all cases, and therefore probably only partially reflect the true scale of the issue. Of these, the over-60 age group comprised 31% of cases, and most died of natural causes. Therefore, death is an issue which the visitor economy needs to be prepared for, especially if it is interacting with an ageing population. Although we will return to ageing and mortality at different points in the chapter, an engaging discussion by Filimonau and Brown (2018: 69) introduced the concept of 'last hospitality' in terms of death and the role of the funeral-care sector in the visitor economy, and observed that 'the current lack of attention to funeral services in consumer research can be partially explained by the general public's unwillingness to face and discuss the topics of death and mortality'. Even so, studies such as Bering *et al.* (2017) illustrate the significance for the sector, particularly in the case of guest deaths in hotels.

Physical activity is a very prominent theme in leisure research and ageing, since it connects leisure and well-being in terms of helping to slow down the progress of some degenerative diseases in older age. The WHO (2010) global recommendations on physical activity for older adults over the age of 65 years include walking, dancing, gardening, hiking, swimming, as well as transportation/mobility (i.e. walking or cycling) and occupational (activity where an individual is still working). The recommendations also spanned household chores, play, games, sports or planned exercise, in the context of daily, family, and community activities. This assessment moves away from a narrow research focus on ageing and sport (Jenkin *et al.* 2017) and activity to target specific diseases (e.g. Laukkanen *et al.* 2011). The purpose of such activity, as WHO (2010) outlined, was to develop cardiorespiratory and muscular fitness, to contribute to bone and functional health, and to further reduce the risk of depression and cognitive decline such as dementia. In terms of the recommended amounts of physical activity, this was seen as consisting of at least 150 minutes a week of moderate aerobic activity or 75 minutes of more vigorous activity. This guidance was also adapted for people with mobility restrictions and to help reduce the risk of falls. Where health conditions were more limiting, then the advice was to remain as physically active as possible. For the visitor economy, engaging people out-of-home in tourism and leisure activity has clear well-being benefits. But this may well require businesses to change the nature of the products, experiences and services they provide so they become more age friendly. Likely changes will involve understanding the interests and

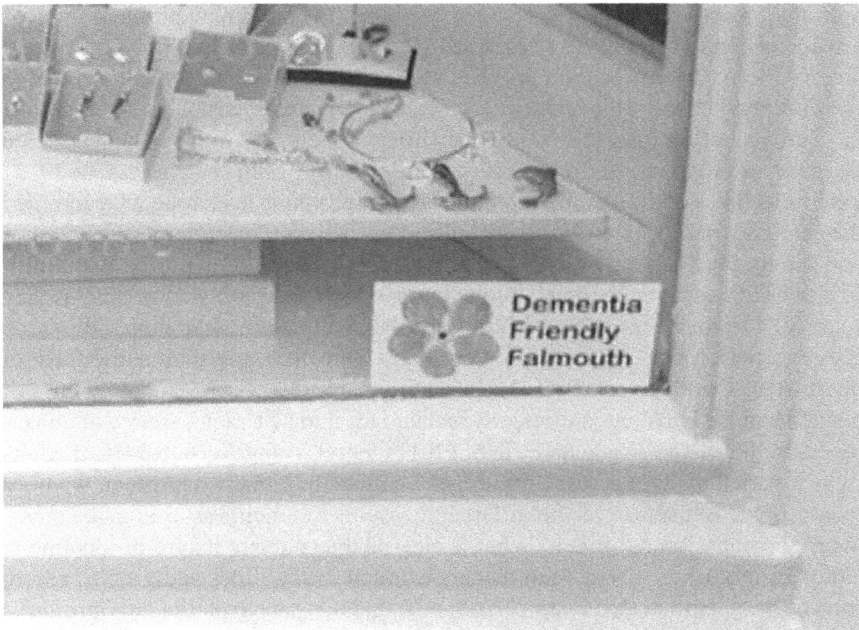

Plate 5.1 A window sticker indicates that this business is a member of the town's dementia-friendly scheme

needs of the market and the type of adaptations and accessibility adjustments required. Probably the one sector of the visitor economy that has developed this focus for many years is the cruising sector, which has a strong affiliation with the over-65-year-old market (see Chapter 1). There is also a growing recognition that as analysis of ageing becomes more visible in the thinking of the visitor economy, it will need to embrace the diversity of the market, paying attention to the different stages of ageing, the life course and issues of diversity and inclusivity. For example, Kalyani *et al.*'s (2015) analysis of diabetes, ageing and race found that race and gender had a very important bearing on experiences of leisure in terms of the time and resources available to pursue leisure-time opportunities. Yet one of the most prominent current issues associated with ageing is loneliness.

Loneliness and ageing: a problem for society or an opportunity for the visitor economy?

Understanding loneliness as a societal issue

Loneliness is a major issue among older groups in society (and also in the general population); this is becoming increasingly evident in advanced developed societies, as more and more people are living alone and reaching much older ages than would previously have been the case. It is not a new problem in modern society, but Weiss (1975) described it as the most prevalent form of distress among the American population. The scale of loneliness as a societal issue is demonstrated by the wide range of social surveys which focus on it. For example, the European Commission's 2018 policy brief 'Loneliness – an Unequally Shared Burden in Europe' (https://ec.europa.eu/jrc/sites/jrcsh/files/fairness_pb2018_loneliness_jrc_i1.pdf) identified around 30 million adults in Europe who admitted they frequently felt lonely, with 75 million people only meeting up with friends and family on a monthly basis. In the UK, Age UK (2011) reported that a million over-65s suffered feelings of loneliness, which was hardly surprising as they spent 80% of their time at home. Similarly, the Institute of Public Policy Research (2009) reported that 2.4 million older adults were suffering with some form of depression. Cacioppo and Patrick (2008: 5) indicated that approximately 20% of residents in the USA felt 'sufficiently isolated for it to be a major source of unhappiness in their lives'. Yang and Victor (2011) found a range of 6 to 34% of adults (which varied by age) across 25 European nations felt lonely. However, we need to add a note of caution, as the different measures of loneliness used make cross-country comparisons difficult. What is clear is the endemic nature of loneliness; although it is often stereotyped as concentrated in the older age groups, recent research (e.g. British Red Cross 2020) has found it has been particularly profound among younger age groups (e.g. teenagers and young adults) during the Covid-19 pandemic. In our interviews with respondents, one representative of an age charity (R1) wondered whether having computer or telephone skills made it easier for people to deal with loneliness, and concluded it did not, as in-person friendships and social interaction were more important:

> the thing [about] … friends, so Facebook friends you can have 700 but you don't have necessarily connection with all those people so it's about going back to the three meaningful things [you must have something meaningful to do, you must have someone to love and care for or something to love and care for], so it's about quality of relationship not quantity thereof. So on a superficial level being connected to lots of people matters but I believe in the magic of three. So for our XXXXX project we always aim to make people make three quality connections, so we try to make sure that people have like three friends, you know, [including] the older adult who has no one, … there's not really [any] science behind it, it's just as a social worker I have observed that three is the optimum number in a friendship, little groups of three work particularly well, two best friends is better than one best friend, you're not overly dependent. So three seems to work extremely well [and] … because of that, you know, we work really hard at sort of making those sort of connections with people and promoting them and … guiding them, you know, we provide free opportunities for people to get together, have exercise, have coffee, things like that where they can make their own friendships but in terms of people being able to weather the storms of life, three good friends really works well as a group.

But what exactly is loneliness as a construct? And why is it important in terms of the visitor economy?

From a social-psychological perspective, studies such as Perlman and Peplau (1981) demonstrate that people who are lonely typically look to become socially connected with others and to seek affirmation from them in their relationships, which helps to address feelings of loneliness. As Gössling *et al.* (2018: 1586) argued, 'there is considerable evidence in the literature that individuals make huge efforts to socially belong, and that, indeed, there is a "universal need to belong" (Heinrich and Gullone 2006: 696)'. Other studies such as Toepoel (2013) emphasise the importance of the association between the quality and quantity of social interactions one has, and the need to maintain and develop these, especially in older age, as the quote from R1 above highlights. These connections perform a key role to help act as a buffer against major life events such as retirement, death of kin and friends, and failing health. This concept of connectedness is central to loneliness and Gössling *et al.* (2018) provide a convincing argument that travel for pleasure may also fulfil this connectedness function when it is about strengthening family ties; that is, visiting friends and relatives. This helps offset the psychological effects of loneliness. Increasingly such studies are also demonstrating the way in which the digital age offers alternative ways to remain connected. As Gössling *et al.* (2018: 1590) point out: 'VFR tourism, along with volunteer tourism, independent travel where backpackers create communities … and virtual travel (including online forums) are examples of the many manifold types of travel that tourists may undertake in search of connectedness'. These actions may help to create a community of people, a process enabled by digital technology (Firth and Mellor 2007).

Loneliness as a social-psychological construct

As Perlman and Peplau (1981: 31) poignantly put it, 'loneliness is a subjective, unwelcome feeling of lack or loss of companionship' as a state of mind. Loneliness is often described as social isolation, but as Dykstra (2009: 92) indicates, 'it is important to draw the distinction between loneliness and social isolation … Loneliness is a negative, subjective experience, whereas social isolation is the objective condition of not having ties with others'. Weiss (1975) differentiated loneliness into emotional and social loneliness, and the two categories have been described as follows:

> emotional loneliness is missing an intimate attachment, such as a marital partner, and is accompanied by feelings of desolation and insecurity, and of not having someone to turn to. Social loneliness is lacking a wider circle of friends and acquaintances that can provide a sense of belonging, of companionship and of being a member of a community. (Dykstra 2009: 92)

In seeking to explain loneliness as a construct, Dykstra (2009) identified three sets of factors that can contribute to loneliness in older adults:

(a) The quality and number of social networks an individual has, which often distinguishes between the deeper, less lonely experiences of older married adults compared to those of single older adults.
(b) The desire and nature of social relationships an individual has as well as the quality of these relationships. Dykstra (2009) suggested that this may explain why those who are socially isolated may not necessarily feel lonely if the quality of their relationships is deep and meaningful.
(c) Predisposing conditions may mediate or impact (a) and (b), as poor social skills may have an adverse impact on the two former set of factors as well as leading to a lack of confidence in forming relationships.

Dykstra (2009) posited that becoming old is often equated with becoming lonely, and that this is a commonly held stereotype. Drawing upon international studies of loneliness, Dykstra (2009: 97) concluded that 'loneliness is common only among the very old, that is, those aged 80 and over. Loneliness levels show little variation across midlife and early old age, whereas young adulthood is characterized by a relatively high prevalence of loneliness'. What is clear from this assessment is that the conventional imagery of ageing and loneliness being synonymous is not completely borne out in the research evidence. Instead, we need to seek a better understanding of the causes and impacts of loneliness, and the interconnections between it and ageing. Nonetheless, this does not necessarily mean that loneliness in old age is not an issue.

Evolving research agendas on ageing and loneliness

O'Connell (2017: n.p.) describes loneliness as a taboo subject, as the following emotive narrative poignantly demonstrates:

> loneliness can be a silent assassin that wreaks havoc on people's lives – and experts want it to be a health-policy issue … It can be transient, a passing state precipitated by unfortunate events, or it can cradle you in its grip for many years … [it may appear as] … sadness that comes from lacking friends or company … it is the searing, destructive pain of finding yourself involuntarily unable to connect with other people that wreaks havoc on everything from your mental health to your immune system … A growing body of evidence links loneliness to dementia, depression and accidents, disrupted sleep patterns, altered immune systems, higher levels of stress hormones, and inflammation.

This excellent summary of the personal effects of loneliness shows that loneliness is a deep-seated issue when it impacts an individual, and has profound, but often hidden health implications. Changing family and social structures have led to a debate on whether loneliness is on the increase in society, particularly in an ageing population, which is affected by these factors as well as the impact of later-life events (e.g. bereavement). Murthy (2020) has described a global loneliness epidemic that has been accelerated by Covid-19. One way to contextualise this debate is to examine the evolution of this area of study within ageing research and the way this issue is being debated in the new millennium.

According to Shanas (1971), from a sociological perspective, concern with ageing and loneliness can be traced to the 1940s and landmark works in the postwar period in America (e.g. Burgess 1959; Cavan *et al.* 1949). According to Shanas (1971: 160–1) in charting the historical analysis of loneliness research and the formative influences,

> The most important book on the family in old age published before 1965 is undoubtedly Peter Townsend's The Family Life of Old People (1957) … [which] … made an intensive study of old people living in Bethnal Green, a working-class area near central London. His findings, like those of Streib and Thompson [1958], indicated that old people were not isolated from their families, and that in old age family life became not less important but more important. Further, he found that only a small minority of old people could be described as isolated, and in a brilliant analysis, he distinguished between isolation, or seclusion from others; desolation, or social loss; and loneliness, a subjective rather than an objective state of being.

A good follow-on study that revisits and updates many of the classic studies is Phillipson *et al.* (2001). This study analysed the impacts of changes in household composition, the spatial proximity of family, kin and relatives, the interactions and help provided by close family, contact with neighbours, social relationships, involvement in leisure activities and relationships with friends. This well-grounded sociological study helps us to understand some of the influential factors that have shaped the experiences of loneliness among an ageing population in the post-war period.

If we move 64 years on from Townsend's (1957) landmark study and reflect on the assessment of the same issues by Age UK (2018: NP), we see that while loneliness is a common human emotion, it has considerable resonance with research on ageing. Age UK provide an all-embracing and empirically verified account of the situation in the UK that is probably not dissimilar to the situation in many other advanced developed nations:

> Loneliness is a negative feeling people experience when the relationships they have do not match up to those they would like to have. When this feeling persists it can have a negative impact on well-being and quality of life. Loneliness often begins when people lose significant relationships or the opportunities to engage in ways they find meaningful. People aged 50 and over are more likely to be lonely if they do not have someone to open up to, are widowed, are in poor health, are unable to do the things they want, feel that they do not belong in their neighbourhood or live alone. The proportion of people aged 50 and over living in England who say they are often lonely has remained similar for at least a decade. If we do not tackle this issue ... the number of older people who are often lonely will increase to 2 million by 2026. The reasons people feel lonely are personal, so the support needed to help them cope with or overcome these feelings must also be personal. Social activities are an essential component of successful approaches to tackling loneliness, but for many lonely people such activities are only effective when complemented by emotional and practical support to access them. (Age UK 2018: n.p.)

Yet Dykstra's (2009) review of the international studies on loneliness provided a different perspective:

> [the] commonly held assumption is that loneliness has increased over time. Findings showed the opposite trend: levels of loneliness have been decreasing over time, albeit slightly, or they have remained unchanged, depending on the studies that are considered. Loneliness is not the only outcome showing a change for the better. In so far as they are available, trend data reveal that since the 1950s average happiness has increased slightly in rich nations and considerably in developing nations.
>
> (Veenhoven and Hagerty 2006)

There is considerable divergence as to the extent of loneliness and whether it is increasing or decreasing as a societal trend, particularly in relation to its association with ageing. Phillipson *et al.* (2001) drew attention to the major societal changes that have occurred that impact upon the issue of loneliness among an ageing population. They argue that the way we connect with people is changing as family relationships change: people now have personal communities populated by friends, which can be as significant as kin and relatives. Gössling *et al.* (2018: 1586) conclude that 'late modernity in developed nations is charac-terised by changing social and psychological conditions, including individualisa-tion, processes of competition and loneliness. Remaining socially connected in this situation is becoming increasingly important'. This is particularly pertinent in older age, especially as the digital revolution may have bypassed some older citi-zens, as noted in Chapter 2. Where loneliness exists in older age, Dykstra (2009) argued that a common route to break the cycle of loneliness is greater engagement in leisure-time activities to reduce one's sense of isolation. Scharf and Keating (2012: 10) suggested that 'population ageing has not resulted in the weakening of family ties but signifies a changing balance between older and younger people in society' in relation to intragenerational families. This produces a more complex pattern of social relationships than those depicted in the classic studies of society and life dating from the 1950s. Yet, where there is an underlying concern about ageing and loneliness, leisure activities have a key role to play, and this creates specific opportunities for the visitor economy in its broadest sense, ranging from entertainment, hospitality, and events to performing arts and travel.

Leisure, tourism and the visitor economy and loneliness

Toepoel (2013: 355–6) summarises the important role that the visitor economy and the different environments in which leisure activity occurs might play in addressing loneliness in old age by maintaining social integration. However, Toe-poel (2013) found that age was negatively related to the size of older people's networks and their social proximity to network members, concluding that age was positively correlated with experiences of loneliness. Theoretical explanations of the role of leisure in older age (see Chapter 2) may begin to offer some explana-tions of how loneliness may become a problem in older age, as people enter retire-ment and potentially withdraw from the employment sphere, children leave home and peers begin to die. Social disengagement theory has been used to depict the impact of these life events, but Toepoel (2013) draws upon studies such as Pet-tigrew (2007) to argue that individuals can take control of the issue of loneliness. As continuity theory has indicated, formal and informal leisure can help ageing people to remain socially active and thereby reduce loneliness. Leisure activities that reduce social isolation include: volunteering (Cornwell *et al.* 2008), sports (Liu 2009), and the use of Internet (Firth and Mellor 2007). There is also growing evidence that interventions by the hospitality sector are also having a beneficial effect. For example, some cafes have established schemes such as coffee and chat or the chatty café (https://thechattycafescheme.co.uk/) (sometimes called chat and natter) where people can sit at a table which is designated for those who are will-ing to kick-start conversations. There are over 400 such cafes in the UK, and while the word loneliness is not used to promote them due to its negative connotations, their underlying motive is to help with this problem by stimulating conversation.

This evidence indicates that many leisure activities may have a specific or pri-mary social outcome (e.g. meeting up with friends or seeking opportunities for social engagement), or this aim may be met indirectly through group activities (e.g. a secondary outcome such as going to a concert) where social relationships are renewed or strengthened. In terms of tourism, Gössling *et al.* (2018: 1596) suggested that:

> mobility is evermore relied upon as a means of social connectedness, [and] further growth in tourism can be expected in both the developing and developed world. This may provide further opportunities for the tourism and transport industries to capitalise on feelings of loneliness and connectedness through advertising campaigns, but it also presents challenges for the sustainability of the sector. Restricting mobility for environmental reasons will encounter considerable resistance due to the threat it would place on self-identities and the opportunity for social relations.

A range of studies on loneliness and tourism (e.g. Farmaki and Stergiou 2019) show that holidays have a role in reducing feelings of loneliness, and as Pagán (2020: 1394) identified, this was most notably the case in the 40–64 age group. Numerous holiday companies specialise in the older age groups and in providing accompanied trips for single people to encourage travel. Laterlife.com highlights opportunities for single travellers (see www.laterlife.com/retirement-c5/retirement-holidaying-alone.htm; the company facilitates an estimated 5 million trips a year in the UK) as well as leisure-specific opportunities to grow one's social connections, recognising the self-help approach to loneliness. Other advice can be garnered from Ideas For Over-50s, Singles Holidays & Over 60s Solo Travellers (www.silvertraveladvisor.com) and TripAdvisor.

Coronavirus and loneliness

The coronavirus pandemic is arguably a crisis that has exacerbated the prevalence of loneliness for those people who have been forced to spend longer periods of time alone; this has been compounded by social distancing and less contact with family and friends. Among over-75-year-olds, who are deemed to be a high-risk group, shielding strategies have accentuated social isolation and loneliness, as many studies in the media have demonstrated (e.g. O'Shea's 2020 analysis of Ireland's longitudinal study of ageing is fairly representative of results reported in other countries). O'Shea (2020: n.p.) found that 'increased feelings of loneliness, anxiety and isolation' were being reported during the pandemic by greater proportions of older respondents, although studies invariably report the loneliest group during lockdown were the 18–24-year olds (Gen Z). Even so, O'Shea noted that over 2 million people over 60 years of age were estimated to be shielding from coronavirus in mid-2020. Covid-19 self-reported mental health worsened in the UK during its first lockdown, with the main symptoms being a rise in anxiety but no major increases in depression. The lockdown most affected those with pre-existing conditions of anxiety, depression and loneliness. *The Elder* (2020) found that during lockdown, 20% of those aged 70 or over were in contact with friends and family less than once every 14 days; 28% of over-70s were not confident using technology, but over 50% said they were having to use it more frequently; and 10% stated they were more prone to suffer anxiety or depression during lockdown. Other commentators have described the impact of the pandemic as causing a loneliness epidemic (Ducarme 2020; Murthy 2020); social isolation can also increase the risk of developing dementia or exacerbate the symptoms, leading to what some geriatricians have described as an increased risk of 'premature mortality' (Klein 2020). In terms of the future of loneliness, Friends of the Elderly (2014) reported that trends to 2030 may well see more single older people in the 60–75 age group, a greater proportion of men than women in the older population than previously and a demand for more accessible solutions that involve intergenerational families (which could be interpreted as a greater demand for leisure and tourism among these groups). They also forecast a growth in out-of-home and in-home socialising, with as more fluidity in older people's leisure time, supported by the use of technology to monitor the well-being of ageing people in their homes. Even so, loneliness is often described as a mental health issue, which is explored in the next section.

Mental health and well-being

Mental health is a widely used construct, particularly in relation to ageing, but rarely is it referred to in the popular literature and media in terms of its constituent parts; instead, it gets used as a general catch-all term that oversimplifies its complexity. Mental health comprises an amalgam of emotional, psychological and social attributes that shape our well-being. In simple terms, our mental health impacts upon how we feel, think, and behave and pm the decisions we make about relationships and interactions with other people. It also affects how we deal with stress, anxiety and life. Mental health is complex because there may be physical, emotional and psychological reactions to different stresses that may coalesce to cause mental illness (e.g. environmental factors, genetic issues, biochemical imbalances in the brain). Among the common traits that medical professionals associate with the onset of mental illness are: variability in a person's moods, including mood swings; increased and long-term sadness or irritability; high levels of worry and anxiety; social withdrawal; and changes in sleeping habits.

Mental illness is one of the great hidden problems in modern society, given its widespread prevalence. In the USA, estimates of mental illness suggest 25% of adults aged 16 or older experience some type of mental illness in any year. The organisation Health Poverty Action Group (www.healthpovertyaction.org) states that 300 million people worldwide suffer from depression and that 1:4 people may suffer from a mental illness in their lifetime. In the UK, 1:6 adults experience mental illness in their lifetime, while globally the three most commonly cited facets of mental illness are:

- *Anxiety disorders*: these are treatable conditions, although the Anxiety and Depression Association of America suggests that only 37% of those affected seek treatment.
- *Major depressive conditions*: these span persistent depressive disorder, major depression, seasonally affected disorder, psychotic depression, situational depression and other types.

- *Bipolar disorder*, often referred to as manic depression with major swings in moods, called mood episodes: i.e. ranging from extreme happiness (mania) to more melancholic episodes (a depressive condition) which can be triggered by periods of high stress (e.g. bereavement, trauma, drug, alcohol or substance abuse).

Other mental health problems may include eating disorders, Obsessive Compulsive Disorder (OCD), personality disorders, post-traumatic stress disorders and psychosis. Mental illness is far more prevalent among the over-60 age group, with the World Health Organization estimating that, on a global scale, 15% of this group experience mental illness every year. This is compounded by the prevalence in this group of severe cognitive impairment – dementia (which we will return to later in the chapter); this affects around 10% of over-65s.

At a global scale, mental illness is not adequately diagnosed because it has a social stigma attached to it which makes people reluctant to seek help or advice. As Birren *et al.* (1992) showed, the relationship between mental health and ageing is complex and has begun to attract a greater degree of attention from the medical profession (see Hantke *et al.* 2020). This has been accompanied by a greater awareness at government level, leading to the development of health policies to target mental illness. Among the proponents of tackling mental illness, Clark *et al.* (2018) identified the cost-benefit this would provide in the form of major financial savings for governments. The Foreign and Commonwealth Office (FCO 2013), citing the National Health Service, described factors that impact tourists' mental health when travelling, illustrates the significance of this neglected area of travel medicine. For example, in extreme cases, psychiatric problems may cause the repatriation of tourists (Streltzer 1979; Felkai *et al.* 2020). Seeman (2016), Felkai and Kurimay (2017), Valk (2017), Airault and Valk (2018) and Marcolongo *et al.* (2019) also drew attention to the neglect of this aspect, as travel medicine has predominantly focused on infectious diseases affecting older travellers (e.g. faecal contamination and river cruises, Jang *et al.* 2019), routine problems such as seasickness (e.g. Wilson 2017), and health concerns when flying (see Bekö *et al.* 2015; Low and Chan 2002; Iqbal *et al.* 2003 on venous thromboembolism). The FCO (2013) highlighted that there was a much greater need for attention to the potential mental health issues facing tourists. The FCO (2013: n.p.) referred to guidance from NHS Scotland (n.d.), which summarised the risk factors as comprising:

- *   *Separation from family and friends,*
- *   *Time zone changes and jet lag/sleep deprivation.*
- *   *Disruption of normal routines and travel delays.*
- *   *Unfamiliar surroundings and presence of strangers.*
- *   *Culture shock and sense of isolation.*
- *   *Language barriers.*
- *   *Use of drugs and alcohol.*
- *   *Physical ill health during travel.*

* *Forgetting to take medication regularly.*
* *Type of travel; some forms have a higher risk e.g. business, family events (wedding/funeral) and volunteer/aid work.*

This was part of a FCO campaign supported by the Mental Health Foundation, whose #travelaware campaign offered guidance on travelling abroad and the likely effects on mental health. Some studies have also pointed to the positive benefits of domestic travel for stress reduction among the elderly population (e.g. Chang 2014) although many of the travel medicine studies have focused on risk factors associated with the physiological and psychological stresses of travel (e.g. Leggat 2005), which reiterates the importance of pre-travel advice for more ageing travellers (e.g. Sanford 2002). Within the remit of mental health, depression is one of the most prevalent problems that travellers, especially the elderly traveller, may face. Clark *et al.* (2018) suggest that depression is 50% more disabling than other physical illnesses such as asthma, arthritis, diabetes and angina, all of which are common diseases within the ageing population.

Depression, anxiety and well-being

Depression is the leading cause of disability at a global scale, affecting over 121 million people (Nimrod *et al.* 2012), with the WHO (2020) now estimating that this now affects over 264 million people. It is the leading cause of mental health illness (Vos *et al.* 2013). Beutel *et al.* (2019) illustrated that in terms of the onset of depression in old age, key predictors were: loneliness, generalised anxiety, social phobia, panic, personality, smoking and cancer. Among an ageing population, when feelings of loneliness combine with other factors, this can in extreme cases lead to suicide, the risk of which is particularly heightened after the loss of someone close. Fortunately, depression may be treated with drugs and psychotherapy, as well as 'self-help' techniques which can involve the visitor economy environments, with green open spaces and blue spaces seen as particularly therapeutic for the condition. Nimrod *et al.* (2012: 420) argued that depression is characterised by *anhedonia*, the inability to derive pleasure from normally pleasurable activities (i.e. the absence of the interest and enjoyment so commonly associated with leisure). Some scientists have also argued that there are biological causes of depression that emanate from a lack of serotonin and norepinephrine in the body: these are neurotransmitters in the brain that seem to have a connection to depression, which is then treated using medication. Hence both cognitive and biological factors may be involved in this condition. According to the Mental Health Foundation in the UK, citing Ferrari *et al.* (2010), depression is the main driver of disability worldwide.

The WHO (2020) suggested that more women than men are affected by depression and that it affects all age groups. Depression raises issues related to positive psychology and the importance of happiness to well-being. Depression and the emotion of happiness (see Layard and Clark 2015) feature in the domain of subjective well-being (SWB), which is concerned with notions such as living well and the pursuit of happiness. McCabe and Johnson (2013: 44) argue that

> happiness has been recognised as an important goal of society and there has been an explosion of research undertaken in terms of understanding what makes people happy ... Happiness is sometimes more broadly defined as SWB, since improvements in objective circumstances have proven to yield limited increases in happiness ... Similarly of interest is how some people remain chronically happy despite personal tragedy, whilst others perceive themselves as unhappy though surrounded by comfort.

As Nimrod *et al.* (2012: 421) indicated, depression is an umbrella term that covers a wide spectrum of conditions ranging from the less severe (e.g. abnormal thoughts, emotions, phobias and personality disorders) through to psychosis, schizophrenia and suicide. In extreme cases, it has a devastating impact and leads to suicide; it is estimated to be responsible for over 850,000 deaths worldwide annually (Nimrod *et al.* 2012). De Leon *et al.* (2001) demonstrated that each suicide impacts upon at least 5–6 other people, and up until the new millennium it was a leading cause of death among over-64-year-olds; however, this has since been counterbalanced as more younger people are dying from suicide. The ratio of female to male suicide rates is at around 1: 3.6, though rates vary by country and culture as de Leon *et al.* (2001) showed and are also affected by the fact that suicide is not always recorded as the cause of death. As de Leon *et al.* (2001: 12) argued

> One phenomenon which has not yet received sufficient attention is the proportional increase in suicide rates in old age, particularly in Latin countries, with an almost parallel decrease in Anglo-Saxon ones ... A similar trend has also been observed in South American countries ... The explanation for this is probably related to the differences existing between these two cultural poles, over and above all the less favourable recent social changes affecting the elderly in Latin countries ... explained by a breakdown in family structure in recent decades.

Another factor in this is the more comprehensive social welfare provision in many Anglo-Saxon countries. Yilmaz and Karaca (2020) found that dissatisfaction with life combined with an absence of leisure-time activities plus the absence of a partner among an ageing population increases the likelihood of depression and risk of suicide.

Depression typically affects 1:4 people in UK in any year, and similar proportions of people can be identified in other countries where studies exist. Lee *et al.* (2018: 654) acknowledged the absence of studies on ageing and depression, and highlighted that 13–21% of older adults live with depression in Taiwan, a trend which is on the increase. It is also noted by Lee *et al.* that older adults had the highest suicide rates, with the main cause being depression. Furthermore, loneliness is a major contributor to depression because it has a negative effect on SWB. Other key factors, such as the sedentary behaviour that characterises older age, are seen to increase the risk of depression and suicide. Although some studies have adopted a life-course approach and found that adults do not stay constantly depressed across the entire life course, other studies focus on the overall impact of depression. As Sjöberg (2018: 51) attests, 'in older adults, depression is one of the most prevalent mental disorders, and a common cause of reduced life-satisfaction and functional impairment'. Yet there are self-help tools that can help with depressive symptoms. Research suggests that positive emotions lead to a variety of effects that strengthen general psychological capacity. Joy especially seems to make one more resilient. SWB and happiness may be connected to the various activities people undertake in their leisure time, and so it is important to explore the leisure–depression relationship. In fact, there is evidence to suggest that laughter is a significant contributor to alleviating depression among an ageing population as it helps to improve one's mental well-being and immune system by decreasing levels of cortisol. Laughter is also proven to alter dopamine and serotonin levels in the body because it may lead to endorphins being secreted that help people to improve their depressed mood (Yim 2016).

The relationship between leisure and depression has seen a significant impetus in research from positive psychology focusing on well-being. As Filep and Laing (2018: 344) argue, '*positive psychology – the study of what makes life worth living – has been precisely defined as an area of study that seeks to highlight the role of positive emotions, character strengths, and positive institutions serving human well-being and happiness*', and their study places particular emphasis on how individuals may use their leisure time to address the emotional and debilitating effects of depression. Filep and Laing (2018) trace the roots of positive psychology to Ancient Greece and the work of Aristotle and Plato, with the key concept being 'eudaimonia (daimon meaning "the true self")'. Eudaimonia has been defined as a higher state of flourishing. It is largely created through self-development and self-realisation by individuals (Filep and Laing 2018: 344). In order to achieve eudemonia, an individual will strive towards excellence and achieving their potential. In juxtaposition to eudaimonia is hedonia, where the individual's focus is on happiness as a form of pleasure (Hartwell *et al.* 2018).

Lee *et al.*'s (2018) review of international studies on depression noted that leisure activities were important protective factors against depression, and that it was important to have a diversity of leisure activities such as meeting people or group activities. Yet, physical activity was not a universal panacea for decreasing depression, with studies showing a more positive outcome for women compared to men. Nevertheless, engagement in leisure activities did illustrate their positive benefits in reducing depression. This supports the positive benefits of leisure activity for protecting one's well-being, obviating the impact of stress or stressful events and enhancing resilience in coping with these events. As Nimrod *et al.* (2012: 421) illustrate, leisure participation has distinct social-psychological benefits in commercialised and non-commercialised settings:

> on coping with stressful life events … leisure participation facilitates coping with such events in two ways: (a) leisure that is highly social in nature can facilitate the development of companionship and friendship and,

consequently, social support. (b) the sense of control and competence that leisure activities may generate are important to enduring beliefs of self-determination that makes stress more bearable. These perceptions of social support and self-determination are described as buffers against life stress and when stress is high, leisure's contribution to health is expected to be greater. Arguably, depression would be less likely to follow in those cases as a result. Leisure involvement is then protective in this way[.]

Although this assessment may be debated in terms of its application to individuals and the particular nuances surrounding their experiences of depression, the underlying argument here is that leisure time, if harnessed in a positive way, can have many positive outcomes around stress reduction, socialising and ultimately in helping to address depression. Leisure activities may have a palliative effect on depression by keeping one's mind busy, diverting attention from negative thoughts and creating positive emotions by engaging in distracting activities. Among older adults, Dupuis and Smale (1995) found that leisure activities that promoted self-expression, creativity and freedom of choice were most likely to help address feelings of depression. Other studies, such as Fullagar (2008), Hong *et al.* (2009) and Janke *et al.* (2008), reaffirmed the positive benefits of leisure activities in reducing the impact of depression among aged people. Leisure activities undertaken on a regular basis may be beneficial but conversely, depression may pose constraints to the positive use of leisure time which may necessitate the use of interventions and strategies to address them. As Nimrod *et al.* (2012) noted, the ways in which people negotiate the constraints which depression poses for leisure needs have to be recognised, in order to achieve a positive contribution to the quality of life among ageing people (also see Iwasaki *et al.* 2010; Fastame *et al.* 2018). But how do holidays feature in the leisure–depression relationship? And what are the implications for the visitor economy?

Milman (1998) undertook one of the initial studies of tourism on ageing travellers' well-being, examining happiness before and after an escorted tour, while Gholipour *et al.* (2016) examined happiness and inbound tourism. Dupuis and Smale (1995) highlighted the greater susceptibility to a decline in psychological state due to the ageing process. Therefore, how do holidays enhance well-being and quality of life? A range of studies have examined the happiness–tourism relationship (e.g. Bimonte and Faralla 2012; Chen *et al.* 2013; Chen and Li 2018; Gholipour *et al.* 2016; Kim *et al.* 2015), which is a complex one, typically approached from a positive psychology paradigm (Filep and Bereded-Samuel 2012). Levi *et al.* (2019) provided one of the few studies of tourism and depression which reviewed competing arguments on tourism and happiness, as follows:

- Tourism may enhance physical health and satisfaction with life and mood, as exemplified by Bimonte and Faralla (2012).
- The process of trip planning, involving mental stress, coordination, travel and unfamiliarity at the destination may involve worry, anxiety, and insecurity, which are feelings related to depression (Larsen *et al.* 2009).

A further dimension observed in travel research is that the positive benefits of a holiday on health and well-being diminish rapidly, especially when people return to work. Yet for an ageing population this was less clear. Despite the limitations of a small sample size, Levi *et al.* (2019: 193) concluded that the impact of a holiday on an individual's depression is personal, and dependent upon many factors. This may reflect the findings in leisure studies on happiness that is related to personality and other human traits (Lu and Hu 2005; McCabe and Johnson 2013). Moal-Ulvoas's (2017) study, again with a limited sample size, illustrated the positive emotions which travel can stimulate (in this case linked to spirituality) as a basis for enhancing one's subjective well-being (also see Filep and Bereded-Samuel 2012: 281). Conversely, McIntosh (2020) reviewed the literature on epilepsy and tourism, and showed that air travel may promote an increase in epileptic seizures, since it disrupts sleep rhythms, oxygen levels and increases stress anxiety. In extreme cases, suicide may occur as a tourism experience. Pratt *et al.* (2019) identified the distinction between assisted or unassisted suicide tourism; the former is a form of medical tourism undertaken to end one's life and die with dignity when a terminal illness becomes unbearable and suicide is illegal in the person's own country of residence (Huxtable 2009). Dignitas in Switzerland is one organisation that facilitates assisted suicide, with Gauthier *et al.* (2015) observing that a proportion of those going to Dignitas for this purpose belonged to older age groups. Hay (2015) recognised that some people chose to die in hotels rather than nursing homes, their own home or hospital, and that this will grow in significance as the population of countries age. Further to this, besides the debates on older people and depression (Djernes 2006; Iwamasa and Hilliard 1999), there are concerns in the visitor economy about the effects of depression on the workforce and its economic ramifications (Conti and Burton 1994), and the relationship between work and depression in this sector (Karatepe and ZagarTizabi 2011). These two perspectives provide some balance to the debate, challenging widely accepted views on the value of holidays and tourism to well-being, especially mental health.

A more critical gerontological approach questions the positive psychology paradigm in tourism (e.g. Nawijn 2016), with its focus on 'how individuals could function even better, instead of only trying to cure those that are not functioning well to begin with' (Nawijn 2016: 151). Nawijn (2016) cites studies that show the impact of holiday taking on depression was only short-term (i.e. Bolier *et al.* 2013), confirming that the positive psychological benefits of holidays and tourism on tourist well-being are perhaps overestimated. This is in sharp contrast to studies by Buckley and Brough (2017), Buckley (2020) and Buckley *et al.* (2021), that show a link between nature tourism and happiness and enhanced mental health. Much of the debate around the relationship between tourism and depression is summarised well by Christou and Simillidon (2020), as shown in Figure 5.1.

Figure 5.1 shows that the causes of depression (melancholy) arise from the individual's background and context in terms of the trip process, the psychological elements and the link to depression (the tourism–melancholy nexus). As Kroesen and Handy (2014), Strauss-Blasche *et al.* (2000) and (2005) and Chen *et al.*

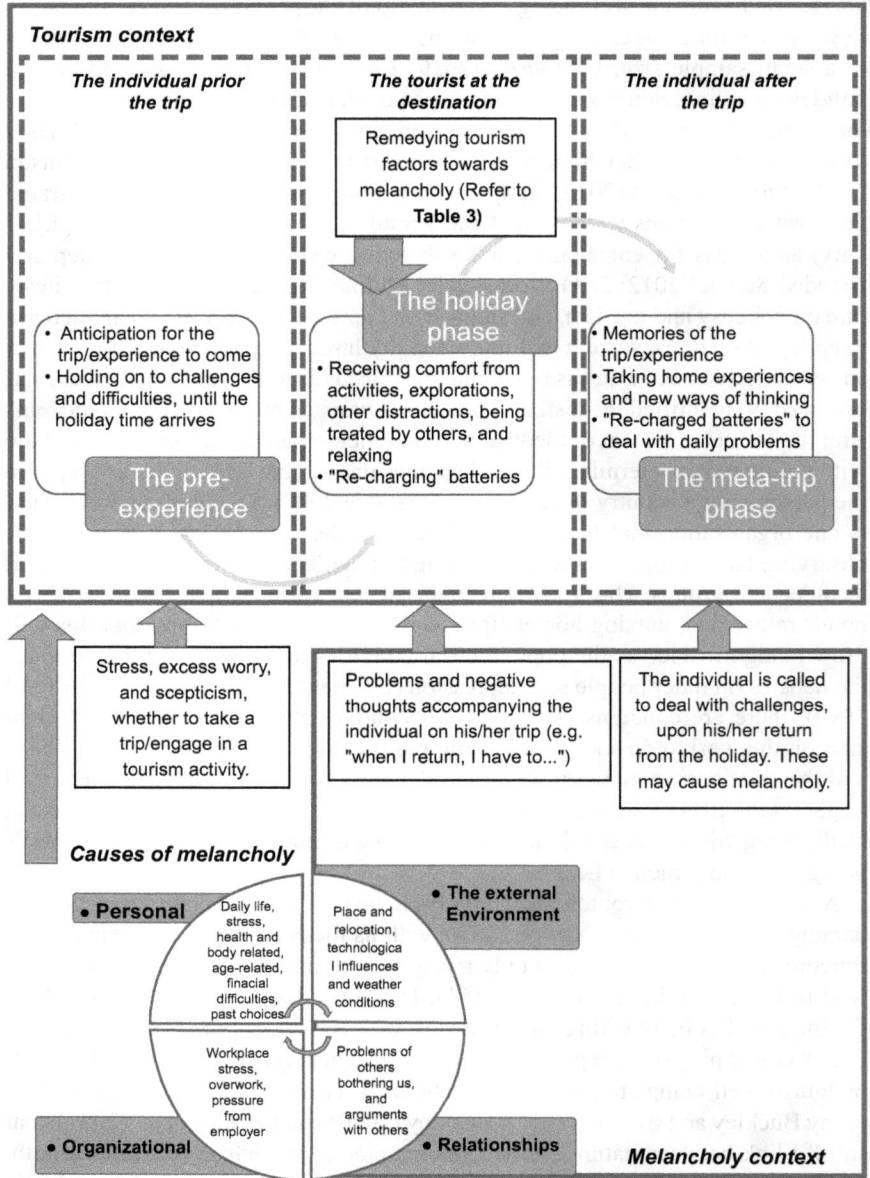

Figure 5.1 Tourism and melancholy nexus

Source: Reprinted from the Journal of Hospitality and Tourism Management, 42, P. Christou and A. Simillidon, Tourist experience: The catalyst role of tourism in comforting melancholy, or not, Page 218, © (2020), with permission from Elsevier.

(2013) conclude, holidays do not raise happiness levels in the long term, a point reaffirmed by Christou and Simillidou (2020).

In 2010, Australia's *No Leave No Life Campaign* promoted taking holidays to enhance well-being (Tourism Australia 2010) and as a means of maintaining good mental health when working. In the UK, making free leisure and tourism activities available on the NHS was trialled to assess how these might impact depression (BBC 2010). Conversely, for the ageing population, holiday periods like Christmas may accentuate social isolation and grieving over lost close companions. Further aspects to be cognisant of include findings from the travel medicine literature that outlines how anxiety can cause panic attacks, triggered by phobias such as fear of flying (aerophobia), claustrophobia from being cooped up in a confined space in an aircraft, or from agoraphobia, a fear of crowded spaces. Ageing tourists and leisure participants are a diverse heterogeneous group, and so it is difficult to draw any generalisations about depression and mental health in relation to travel advice. What is clear is that the greatest benefits for alleviating depression arise from regular, daily or weekly leisure activities and forms of interaction, including laughter, that have been proven to help the ageing population with depression symptoms (Ko and Youn 2011). The evidence on holidays is less certain, which calls the role of leisure-related initiatives and interventions into question. Adams *et al.* (2010) found that the number of depressive symptoms after an intervention on physical activity was not lower, but the intervention helped decrease the number of hours of depressive symptoms (i.e. it had a palliative effect). However, these interventions may protect against developing more severe symptoms (e.g. severe mood disorder) and are important where health conditions lead to a drop in social activities and create a barrier to participation, compounding the depressive symptoms among more aged people (Adams *et al.* 2010). An intervention does help to prevent the development of a vicious circle of decline associated with depression.

Physical health conditions

Cancer

In 2018, 9.6 million globally died of cancer. According to the WHO, cancer is the second most prevalent cause of death worldwide, accounting for 1:6 deaths in 2018. Among the most common cancers in men are: lung, prostate, colorectal, stomach and liver cancers. Among women, the most common cancers are: breast, colorectal, lung, cervical and thyroid cancers. By superimposing the issue of ageing upon global rates of cancer, Sharpless (2018) identified that the convergence of an ageing population and peak cancer incidence occurs between 64 and 75 years of age. Assuming that progress with cancer treatment and prevention continues to be limited, it is thought that the ageing population will lead to a higher incidence of the disease among those over 75 years of age in future. Although it will not be the main cause of death for that cohort (see Palmer *et al.*

2018), Sharpless (2018) nevertheless points to age as a risk factor for the development of cancer. This is because biologically, human cells in an ageing body enter a process of senescence, meaning they no longer divide and grow, which allows cancer cells to multiply and spread. This is also part of a broader process as the body's immune system becomes less efficient in addressing disease and infections. It is this process of ageing that has been the basis for research on cancer-inhibiting treatments and tools that can be used to diagnose the disease early and enhance long-term survival rates. Among older people, treating them for cancer may be complicated by other complex medical conditions, which has created the sub-specialism of geriatric oncology (oncology is the study of cancer and includes medical treatment, such as surgery and radiation therapies).

The leisure literature points to the positive benefits of physical activity in older age, especially exercise that invokes an element of socialising, as it can help with mental and physical health and has a role in cancer prevention. In terms of cancer reduction, Laukkanen *et al.* (2011: 129) suggests that 'the mean intensity of leisure-time physical activity is inversely and independently associated with the risk of premature death from cancer in men'. For women, breast cancer remains a major problem for those over 50, and especially those aged 60–70.

The relationship between leisure and cancer has two distinct perspectives:

- A positive role in that participating in a leisure activity reduces the probability of being affected by cancer in old age (e.g. see Hartung *et al.* 2017) and also works as a palliative measure to reduce the increased risk of depression (Chochinov 2001). Even with advanced cancer, leisure activities still play an important role in quality of life (Waehrens *et al.* 2020; Avancini *et al.* 2020; Kim *et al.* 2021) as well as among cancer survivors (Fitzpatrick 2018).
- A negative effect in that sustained exposure to the sun and other environmental factors contribute to cancer risk in older age, namely skin cancer, which is a notable problem in Australia, for example (Haynes *et al.* 2020), and among tourists who have experienced protracted periods of sunburn (e.g. Weston 1996 and Grandahl *et al.* 2018).

Non-melanoma skin cancer among white-skinned people is the most common form of cancer globally and is increasing in prevalence (Lomas *et al.* 2012), with the highest rates in Australia and south-west England, and the lowest in Africa. In terms of cancer and tourism, the main study on this theme is Hunter-Jones (2005) which illustrated the heterogenous nature of those affected by cancer, with only four elderly respondents in the sample and no clear indication on how cancer affects older travellers. With the onset of cancer, travel habits may decrease, remain the same or increase through the pre-diagnosis, diagnosis, treatment and post-treatment phases. Chung and Simpson (2020) examined the potential value of social tourism initiatives for families caring for people who were terminally ill with late-stage cancer. Social workers and medical staff have a role to play here to highlight the opportunities for social tourism as a form of respite care for affected families and individuals. This draws attention to the relatively undeveloped role of charities in tourism. Turner *et al.* (2001) examined charities' role in relation to fundraising events as a basis for tourism, but the dimension of charities providing social tourism opportunities remains undeveloped. Burki (2019) examined the demand for private cancer treatment in London and the tourism dimensions of this; such tourism generated £378 million for 25 private hospitals in 2017, although the link to ageing was not discussed.

Heart issues

People over 65 are more likely to suffer a heart attack, as blood vessels become less flexible with age and fatty deposits can collect on artery walls, slowing the blood supply from the heart. Factors such as constant exposure to stress and raised levels of adrenaline and cortisol can increase the risk of a heart attack. The effects of sudden cardiac arrest may be fatal, but lesser symptoms may produce less serious effects. Lye and Donnellan (2000) outline the scope and effects of cardiovascular disease, which is one of the leading causes of death among the over-65 age group. Long-term heart conditions such as angina have an impact on around 16% of over-65-year-olds. Blood pressure also increases with age (i.e. hypertension), and this is treated in order to prevent strokes and other coronary events (Lye and Donnellan 2000). Lifestyle facts such as obesity, a family history of heart problems and an inactive sedentary lifestyle may increase the risk of heart problems. Within the UK, Conrad *et al.* (2018), saw a 23% increase in people experiencing heart failure in the period 2002–2014, with diagnosis typically at age 74 for men and 79 for women. Leisure time physical activity (LTPA) is proven to help reduce depression, anxiety, stress and coronary heart disease especially when accompanied with lifestyle changes (e.g. not smoking, reducing alcohol consumption and addressing obesity) (Paganini-Hill 2011; Soares-Miranda *et al.* 2016), although some studies have questioned the efficacy of this (e.g. Li *et al.* 2013). In high temperatures, there may also be a rise in cardiovascular events among the older age groups (Sangkharat *et al.* 2020). Heart problems remain a cause of major in-flight emergencies for airlines among the older age groups, as the following assessment of aviation and flying by Joy (2007) suggests:

> Travel by air most people find stressful to some extent: access to the airport, inconvenient parking, carrying luggage long distances, long delays at check in, security and boarding, jet lag. To this now is added the terrorist threat (it has always been present) and the confiscation of deodorant and nail scissors. And about one–third of travellers also admit to apprehension or fear before and during a flight. Finally, the cabin of a crowded aircraft is not a good environment in which to be taken ill, or render treatment.
>
> (Joy 2007: 1509)

The ageing traveller is also more likely to suffer life-threatening diseases or illness, especially on a long-haul flight (e.g. thrombosis, dehydration) as alveolar oxygen tension rises, humidity is low and people are in cramped seats, while older travellers often have poor physiological reserves. Possick and Barry (2006) indicated that possible contraindications (i.e. conditions that may harm a medical condition) in relation to heart disease include dehydration, which is often experienced on board aircraft. There are also specific heart conditions (e.g. unstable angina, myocardial infarction and having had coronary bypass surgery in the last two weeks) for which air travel will act as a contraindication (a feature we will return to later in relation to travel medicine and the ageing traveller). Wilkinson and Ripley's (2020) analysis of heart disease also highlighted the difficulties such travellers face in terms of getting travel insurance.

Diabetes

Diabetes is a medical condition that is associated with high levels of blood glucose as a result of the body producing insufficient amounts of insulin. There are two types of diabetes:

- Type 1 diabetes, which is where the body is unable to produce any insulin.
- Type 2 diabetes, where the body cannot produce sufficient amounts.

There are also other types of diabetes which are much rarer. In the UK about 1:15 people have diabetes, with 1 million having type 2 diabetes, who have not been diagnosed with the condition. The effects of diabetes can cause additional medical problems for the heart, eyes, feet and kidneys, especially in older people. There are also additional risk factors that can contribute to hypoglycaemia, where the amount of blood glucose drops due to food intake, medical complications and other complex medical conditions. Undiagnosed diabetes can also lead to a higher risk of falls, stroke, confusion and speech difficulties. Older people may also be more prone to hospital admission. For people with diabetes, travel and leisure may pose additional challenges in terms of storing their supply of insulin, and control of glycaemia due to irregularity of meals, stress and time zone changes (Chełmińska and Jaremin 2002). Various guides exist to help people with diabetes when travelling (e.g. Diabetes UK). In terms of leisure, Henze *et al.* (2017) observed that men with diabetes had less involvement with walking and leisure activities and a greater probability of falls and heart attacks.

Obesity

Obesity, defined as the condition of being very overweight with a significant amount of body fat, has been described as a global epidemic. The World Health Organization documented the growth in obesity, estimating that it affected 200 million people in 1990. This had risen to 300 million by 2000 and 650 million by 2016, of whom 115 million suffered with obesity-related health problems (e.g. diabetes, cardiovascular disease, strokes and hypertension). As Small and Harris (2012) illustrated, there was silence on this issue among many airlines. Obesity can increase in older age, where reduced exercise and more sedentary behaviour occurs. There are many other contributory factors to a global obesity epidemic, including increased food portion sizes in developed nations; richer, unbalanced diets based on convenience foods; genetic issues; and a diet high in carbohy-drates and increased levels of alcohol consumption. As the WHO suggests, obe-sity results from an increased intake of energy-dense foods that are high in fat and sugars combined with an increase in sedentary lifestyles, associated with urbani-sation and motorised transport (see Dube *et al.* 2010). As Mezrad (2020) reports, in 2020 around a third of the world's population was overweight or obese, and by 2030 this could increase to 50%. For the visitor economy, Giuffrida (2020) outlined how gondola operators in Venice reduced the number of passengers from 14 to 12 in each vessel due to overweight tourists. In Santorini, Greece, weight limits of 100kg have been imposed for donkey rides to protect the animals from spinal injuries (Newton 2018). In terms of leisure, daily walking and exercise is seen as a solution to the problem, especially owning and walking a dog that forces the owner to exercise. However, some critics also point to the role of all-inclusive holidays and their potential to promote gluttony, as some hotels provide an 'all-you-can-eat' buffet at mealtimes (Koc 2017), which is also an issue for cruise holidays. This remains a neglected area of discussion in the visitor economy.

Falls, accidents and injuries

Falls are an ever-present risk for an older population and there are a multifaceted reasons for their occurrence, ranging from environmental factors like a trip hazard through to medical causes such as a change in medication or biological long-term changes in the body such as muscle weakness, declining vision and health condi-tions (e.g. heart disease, dementia or low blood pressure (hypotension)). In terms of dizziness, two specific conditions that affect an ageing population are: parox-ysmal positional vertigo which medical research describes as a form of vestibular dysfunction; and Ménière's disease, a medical condition affecting the inner ear which may induce dizzy spells (vertigo) as well as hearing loss. According to Lord and Close (2018: 492) 'Falls pose a major threat to the well-being and quality of life of older people. Falls can result in fractures and other injuries, [while] dis-ability and fear can trigger a decline in physical function and loss of autonomy'; the most physically active are less at risk (Cunningham *et al.* 2020). Such an argument reinforces the case for physical activity – healthy ageing interventions, particularly as, according to Nair (2005: 111), falls are a major cause of mortal-ity. According to Wu *et al.* (2020) the annual average age-adjusted falls mortality rate was 38.63 per 1,000,000 population in Australia, and 34.12 per 1,000,000 population in the UK. Their study found that deaths from falls increases with age, particularly among the very old, those over 95 years of age. Wu *et al.* (2020) also found that in the period 2006–2015, fall-related deaths increased among the very elderly. Nair (2005) identified that 44% of men and 65% of women fall inside their residence, and that fall risk rises as the number of risk factors rise.

  Nilsson *et al.* (2015) found that medical conditions did not lead to a reduction in leisure activities, whereas mobility impairments (e.g. fear of falling or need to use a mobility aid) did. Curl *et al.* (2020) reviewed the fear of falling outdoors among older adults, noting the importance of mobility to physical and mental well-being, especially the role of walking and socialising in reducing obesity. In the urban environment, as Chapter 3 illustrated, uneven surfaces, the physical infrastructure and obstructions can all contribute to falls, often classed as 'slips, trips and falls' (STFs). Enhancing leisure and recreational mobility requires the removal of likely fall hazards, since these are a leading cause of hospital admissions (Curl *et al.* 2020). The World Health Organization (WHO 2007b) adopted a three-pillar approach to falls reduction that includes: building awareness of falls prevention across society; improving the identification of the risk factors and their determinants; and developing culturally appropriate interventions.

As a major public health concern, falls are equally important in leisure and tourism settings (Wilks and Page 2003). Falls are often subsumed in tourism and leisure settings as 'accidents', a theme first examined by Page and Meyer (1996), who found that the over-70s age group had a higher accident rate, especially in terms of road traffic accidents (RTAs). These cumulative injuries and accidents impact destination healthcare (Faulkner *et al.* 2020; Page and Meyer 1996), as hospital data shows (Page *et al.* 2001). Dioko and Harrill (2019) examined the impact of death and injuries among ageing travellers, with Larsen *et al.* (2009) recognising that the main causes of worry and anxiety among this group were petty crime and accidents and injuries. In a countrywide study of Finnish tourists' injuries abroad, it was found that 28% of the total injuries were accounted for by people aged over 60, the majority being outpatient cases seeking treatment (Siikamaki *et al.* 2015). This study found that the most common ailments were infections (59%) and injuries (14%). Acute gastroenteritis was one of the most common problems (22.8%), reinforcing Cartwright's (1996) findings. Among the analysis of risk factors for hospitalisation, Siikamaki *et al.* (2015) found that among those who were male and aged 60 or more, vascular disease was a key risk factor. They concluded that age and travel duration explained some of the clusters of health problems, especially in those regions to which more than 50,000 Finns travelled per annum (e.g. the Canary Islands). Other studies (e.g. Evans *et al.* 2001) report gastroenteritis as a common problem (25.7%) among package holidaymakers, but Cossar *et al.* (1990) found that fewer people over 60 years of age reported illness than other age groups, and age was not an important issue in Evans *et al.*'s (2001) study. Even so, Page *et al.* (2005: 381) found that 'what is notable from the scientific studies which are available, is that unintentional injury is a leading cause of tourist morbidity and mortality (e.g. Paixao *et al.* 1991), particularly in environments where recreation is the prime activity focus such as skiing (Hudson, 2000)'.

Degenerative conditions and ageing

According to Tabloski (2004), the majority of older people are women; in terms of health, Deuschl *et al.* (2020) found the highest occurrence of neurodegenerative diseases to be among ageing European men aged 80–84. Jang *et al.* (2018: 3662) proposed that 'the biological basis of human aging remains one of the greatest unanswered scientific questions', and numerous theories and arguments have been developed in biology to investigate what causes the ageing process. Among the agreed aspects of this change are 'physiological changes that take place in the human body leading to senescence, the decline of biological functions and of the ability to adapt to metabolic stress' (Shock 2020). Physiological changes that occur can cause neurodegenerative conditions to develop such as dementia and motor-neurone disease, which contribute to a general decline in brain function (e.g. a decline in memory), although the specific causes of these remain under investigation in scientific research, and a search for treatments and cures continues. One of the most prevalent degenerative conditions associated with ageing is dementia.

Dementia: a silent public health crisis?

Dementia is a brain disease and is not necessarily a natural facet of ageing; there are several forms, the most common of which is Alzheimer's disease, which accounts for between two thirds and three quarters of cases. While most cases of dementia are diagnosed in people over the age of 65, the disease can affect younger people too – this is termed early-onset dementia. According to the WHO (2020: n.p.), dementia is of a

> chronic or progressive nature – in which there is deterioration in cognitive function (i.e. the ability to process thought) beyond what might be expected from normal ageing. It affects memory, thinking, orientation, comprehension, calculation, learning capacity, language, and judgement. Consciousness is not affected. The impairment in cognitive function is commonly accompanied, and occasionally preceded, by deterioration in emotional control, social behaviour, or motivation ... Dementia is one of the major causes of disability and dependency among older people worldwide. It can be overwhelming, not only for the people who have it, but also for their carers and families. There is often a lack of awareness and understanding of dementia, resulting in stigmatization and barriers to diagnosis and care.

Estimates suggest there are 50 million people worldwide living with dementia, with the number of cases growing exponentially, expanding by 10 million additional cases each year. This is a silent public health crisis given its disabling impact and creation of dependency. Estimates suggest that globally, there will be 82 million cases by 2030 and 152 million cases by 2050. In the UK, there are currently over 850,000 people living with dementia, affecting 1 in 6 of over-80-year-olds. The scale of dementia in the UK means that there are 670,000 carers of people with dementia. While this trend in dementia has major implications for health and social care systems (WHO 2020), the economic costs of formal and informal care for people with dementia in 2015 were estimated to be US$818 billion, which equates to 1.1% of global GDP. Conversely, the prevalence of the condition has meant that there is a much greater diversity of travellers who suffer from it; 4% of the UK population are directly affected as carers or people with dementia. This is set to increase in the future as the structure of the population ages. For the visitor economy, this is a visitor segment that, when combined with family and friends, is not an insignificant group, especially as travel behaviour does not necessarily stop in the early stages of the condition. Many people affected by dementia have been the beneficiaries of post-war affluence and increased consumer spending on leisure activity that incorporated travel and tourism. Dementia has now assumed an even greater significance in the UK as a public health issue: the Office for National Statistics (BBC 2016) indicated that dementia (including Alzheimer's disease) had supplanted heart disease as the leading cause of death in England and Wales, accounting for 61,000 deaths in 2016, equivalent to 11.6% of all recorded deaths. In 2020, it remained the chief cause of death among all non-Covid-19-related deaths in England and Wales.

Dementia and the visitor economy: a synergy?

Another reason why dementia remains important for society and the visitor economy is that symptoms connected with cognitive loss may start to interfere with daily life, work and social activities (Steeman *et al.* 2006). Creating opportunities for people with dementia and their carers to engage with a visitor economy that is able to adapt its offers to create leisure time opportunities for them is critical. The argument supporting this is that it can help to maintain a degree of normality in the everyday lives of people with dementia and perform a therapeutic role. From an economic perspective, research indicates that in the UK leisure spending by and on behalf of people with dementia amounted to £11 billion in 2014, and this was estimated to have risen to £16.7 billion in England in 2019; recreation and culture was the top item of expenditure. By 2040, it is estimated that spending on the part of people with dementia and their carers will have doubled (Centre for Economics and Business Research 2019), which illustrates the opportunities this market offers the visitor economy. Page *et al.* (2015) illustrated the implications for destination development, and the demand for places and spaces able to cater to the needs of people living with dementia and their families and carers. Such a focus is clearly linked with the new paradigm of creating age-friendly spaces (also see Chapter 7). Yet the needs of people with dementia and their carers transcend the growing debate on age-friendly cities because of the specific impact of dementia on visitor needs, although there have been attempts to align age-friendly and dementia-friendly initiatives (see Turner and Cannon, 2018). While the UN Initiative focuses on age-friendly cities, it seeks to enhance civic participation and reduce social exclusion. Visitor destinations remain a neglected feature of this debate in relation to dementia, despite the development of evaluation tools to link age-friendly and dementia-friendly city criteria (see Buckner *et al.* 2018).

To date, the contribution of businesses to the concept of living well with dementia remains largely neglected in the business and management literature, particularly in relation to consumer sectors associated with the visitor economy. As a global issue, dementia has seen many countries develop social policies to enhance service provision and dementia diagnosis. To help people with dementia overcome the social isolation associated with exclusion from daily activities (see Genoe and Dupuis 2012; Lokon *et al.* 2016; Eades *et al.* 2016), new approaches to combining business with the solution of societal problems have emerged, as discussed in Chapter 2. There have also been shifts in political thinking that reflect new theoretical approaches to social policy in which businesses have a greater role to play. In the UK, the Prime Minister's Dementia Challenge sought to inspire all businesses to become dementia-friendly (DF) by 2020 (Department of Health 2015). This challenge shows a significant focus on the DF theme, reflected in an international study that ranked the UK as the most dementia-ready country in 2017 because it was seen as a leader in innovative practice (Alzheimer's Disease International and Global Coalition on Ageing 2017). Yet this dementia-readiness is not uniform across all areas of society, and no studies have sought to understand the transformational journey businesses have embarked upon to help address this major societal challenge as part of a civil-society paradigm. The route which most businesses have taken to date is associated with engaging in the development of dementia-friendly communities (DFCs) (see Connell and Page 2019a for more detail on the types of engagement undertaken).

Living well with dementia: a role for the visitor economy in addressing a public health crisis

The prevailing consensus on the long-term care of people with dementia is that it is important to maintain the person's independence and enable them to *live well*. Within the realm of dementia policy and practice, 'is it possible to *live well* with dementia?' is a frequently posed question. One of the aspects of living well is the freedom to go on holiday and enjoy leisure time within the health constraints imposed by dementia, particularly in the onset of its early stage. As Watson (2016: 5) states: 'part of living well with dementia is having fun, in whatever form that takes'. Genoe and Dupuis (2012) identified that one way of living with dementia is to maintain participation in meaningful activity. Interventions that provide stimulating leisure and recreational experiences for people affected by dementia are well established, particularly in the arts and culture arena. Research has shown that individuals with early stages of dementia value outdoor recreation, which provides opportunities for exercise and enjoyment of the countryside. Furthermore, the positive impact on emotional well-being of such recreation increases with both the frequency of outdoor activity and the range of areas visited, appearing to reduce the effect of dementia (Duggan *et al.* 2008). This addresses the perceived 'shrinking world' experienced by those with the disease, as opportunities progressively fall away and taking them up becomes less feasible, shrinking the time–space prism of the individual and carer (see Chapter 3). Paradoxically, increased periods confined to the home not only reduce opportunities for social and environmental encounters that stimulate well-being but can increase feelings of depression. Consequently, being able to continue with a range of activities that were undertaken prior to diagnosis can help people living with dementia to maintain quality of life as far as possible and even mitigate against more rapid deterioration and care intervention. This also has financial benefits for many stretched health systems, helping people to live at home for longer.

Roland and Chappell's (2015) study on meaningful activity for people with dementia from a caregiver's perspective emphasised the opportunities in the early stage of the condition. Being able to go out, interact socially and spend quality time with a partner in a mutually enjoyable pursuit (e.g. on holiday to help stimulate memories of places visited) is particularly valued. Furthermore, Roland and Chappell (2015) emphasised the importance of keeping a sense of normality for those with a dementia diagnosis, often maintained by engaging in activities they used to enjoy and keeping interested in the world around them, which contributes social benefits and supports self-worth. For some, the idea of 'doing things while they can' highlights an added aspect where tourism and leisure experiences can play an important role. However, the reality is that many people with dementia find it difficult to take part in activities that they once enjoyed (Alzheimer's Society 2013; Alzheimer's Society 2015; Alzheimer's Society 2016), and the critical issue is: can the service sector engage more fully to reduce the barriers from a supply-side perspective? To facilitate such meaningful activity as part of a civil society, policy research indicates that DF principles need to be embedded into our communities and business activity. One important development is the creation of dementia-friendly communities (DFCs).

Creating dementia-friendly communities: a role for business?

While DFCs are positioned 'at the forefront of the policy agenda' (Alzheimer's Society/British Standards Institute 2015: v), the aim of establishing such communities was to enable an inclusive locale to develop where people with dementia feel empowered and safe, and have access to services, social networks and meaningful activities. Inherent in this process is the involvement of the business community, so people can have the confidence to use services and enjoy the freedoms associated with an environment that recognises that people have different needs. Essentially, this involves the development of a caring community that supports and accepts social difference (Thomas and Milligan 2015). The Local Government Association (LGA) (2012) and other studies (e.g. Alzheimer's Disease International 2016) have sought to clarify what a DFC (see Figure 5.2) should entail.

As Crampton *et al.* (2012) indicated, individuals are consumers prior to the onset of dementia and remain consumers and users of businesses and services after diagnosis, particularly so in the early stages of the disease. In this respect, nothing changes. However, the role of business engagement in this process remains poorly understood empirically and conceptually, with evidence from large corporations (e.g. Lloyds Bank and its Tea and Talk events in its UK branches) and individual initiatives being limited mainly to ad hoc assessments. Yet as Crampton and Eley (2013: 49) argued,

> the concept of dementia-friendliness is not the exclusive domain of the health and social care world … it is the daily attrition of everyday life where help is

A Dementia-Friendly Community (DFC) is one which:	

o Helps people with dementia to:
 	
  ▪ find their way around easily
 	
  ▪ feel safe when out and about
 	
  ▪ access facilities and services that they are used to (such as banks, shops, cafés, cinemas and post offices as well as health and social care services)
 	
  ▪ continue with everyday tasks in the local area
 	
  ▪ travel around the locality
 	
  ▪ maintain social networks and participation in community activities
	o Promotes education and public awareness of dementia
	
	o Integrates the needs of people with dementia into planning and development
	
	o Empowers people with dementia to contribute to development of appropriate services
	
	o Recognises that people with dementia are a valuable part of the community
	
	o Supports carers and caregivers
	
	o Supports organisations, services and businesses to work towards becoming dementia-friendly
	
	o Brings together stakeholders to establish action plans
	
	o Evaluates actions and progress towards DFC status

Figure 5.2 Defining a dementia-friendly community

Source: Reprinted from Tourism Management, 61, J. Connell, S. J. Page, I. Sheriff and J. Hibbert, Business engagement in a civil society: Transitioning towards a dementia-friendly visitor economy, Page 113, © (2017), with permission from Elsevier.

most needed. People with dementia and family carers find routine activities most difficult – shopping, managing finances, using transport, keeping active – causing them to withdraw and impacting upon their well-being.

Several aspects of DF practice relate specifically to the service business environment, such as the existence of 'respectful and responsive businesses and services' (Department of Health 2015: 10), where staff understand and recognise symptoms of dementia as displayed by customers. In terms of a civil society, this is predicated on mediating institutions that operate within four spheres in society (Edwards 2013; Janoski 1998): the private market (the commercial sector); the public sphere (the third sector and charities); the state (the legislative framework and statutory duties); and the private sphere (family, friends and community associations). As previously mentioned, the PM's Challenge (Department of Health 2015) set out that all businesses should have been encouraged to become DF by 2020 and that industry sectors should have DF charters to guide best practice. In a tourism and leisure context, this is important given that facilitating enjoyable and meaningful experiences through service interventions might contribute to quality of life and well-being, and thereby prolong independence so that those with dementia can continue living at home.

Watson (2016: 5) highlighted that some people are 'reluctant to embrace' DF practices. Despite high-profile media coverage, celebrity endorsements of the 'cause' and an increasing acceptance of dementia as part of our society, lack of awareness and, perhaps worse still, stigma about 'the D-word' (Milne 2010) still persist, and there is a need to permeate the development of DF actions into mainstream society and outside of the medical and care communities. The World Health Organization (WHO 2012) stresses that society must change to becomes more DF (similar to increased awareness of cancer) as part of a civil society. One tool is the role of charities and other advocacy groups working in communities of practice (see Chapter 6) in pushing this agenda forward with the visitor economy. The ways in which new actors can adopt a strong advocacy role through grassroots-level organisations, in order to leverage the support from business and the community to improve people's lives, has become a fruitful area for research (Popple and Redmond 2000). Much of the recent development of grassroots organisations and their role in a civil society to lead a change agenda on dementia is rooted in an emergent academic discourse on social innovation, with its focus on addressing social challenges such as dementia. A rich literature has emerged on social innovation alongside grassroots action at a community level (e.g. see Have and Rubalcaba 2016; Pol and Ville 2009; Gerometta *et al.* 2005; Cajaiba-Santana 2014), and this has a major bearing on how civil society can be harnessed to address societal issues. Even so, the concept of social innovation is not new. As Schumpeter (1909) demonstrated, social value is created through altruistic means by groups and individuals in a capitalist society where a specific cause is championed (also see Moulaert *et al.* 2013). Thus, social value is based on 'mutual interaction and interdependence between individuals' (Schumpeter 1909: 214), where new social practices can meet social needs to further develop a civil society. Moving beyond this literature to understand how this can occur in practice leads us to Porter and Kramer's (2011) notion of creating shared value, mentioned in Chapter 3. One final area of this discussion, our treatment of which synthesises much of the analysis in Chapters 4 and 5, is travel medicine and its implications for the ageing tourist in helping to establish fitness to travel and facilitate participate in tourism; this is the topic to which we now turn.

Travel medicine and an ageing population

Darrat and Flaherty (2019) indicated that older people now represent a greater proportion of overseas travellers than they have since records began, yet Flaherty *et al.* (2018) demonstrated that travel medicine has failed to recognise that the over-60 age group will outnumber the under-60 age group globally by 2050. Darrat and Flaherty's analysis of ageing travellers' attendance at a travel clinic found that 79% had a pre-existing medical condition; in order of prevalence, these were: hypertension, dyslipidemia (i.e. abnormal levels of lipids, which is often associated with high cholesterol), diabetes, insect-bite sensitivity and hypothyroidism. With a growing number of people travelling with diabetes, there will be a greater demand for pre-travel advice, especially among older travellers, around insulin-dose adjustment, self-monitoring, foot care and health management (Jawad and Kalra 2016). As McIntosh (1998) found, there are also key physiological changes between middle and older age that occur, including a decline in cardiopulmonary function, changes in renal function and reduced hydrochloric acid in the intestine, all of which reduce the ability to fight infection. Furthermore, the ageing traveller is more at risk of disease due to a poorer immune system and a higher probability of risk of injuries and accidents. Therefore, as McIntosh (1998) suggests, the ageing traveller is at a much greater risk of infection abroad. The outbreaks of Covid-19 on cruise ships and among returning travellers reiterate what Kain *et al.*'s (2019: 1) findings, that: 'travellers are at high risk for acquiring infections while abroad and potentially bringing these infections back to their home country'. Studies of illness among older adults when travelling (e.g. Gautret *et al.* 2012) have reported that common conditions include: lower-respiratory-tract infections, high-altitude pulmonary oedema, phlebitis (thrombosis), malaria, trauma and injuries, gastrointestinal complaints (e.g. travellers' diarrhoea), among other infections. As we also noted under degenerative conditions, there is a general neurological reduction in the speed at which the ageing traveller can process matters, which can increase their level of stress when travelling. There is also a reduction in human agility and muscle strength, which helps to explain the greater risk of falling and injury when visiting an unfamiliar environment (Bentley and Page 2008).

While travel preparations are an important tool to address and mitigate potential risk (e.g. Lawton and Page 1997), tourists can place seasonal pressure on destination healthcare, reducing the performance level of the care provided (Ezza *et al.* 2019). In the travel medicine literature, travellers report a higher incidence of medical issues in tropical destinations, a feature which Pisutsan *et al.* (2019) confirmed in Asia, with a higher incidence of diarrhoea. However, again in this case the ageing traveller was not specifically identified in the study. There is a well-developed literature on cruising and older people in relation to travel medicine. Oldenburg *et al.*'s (2016) study of shipboard cruise tourist deaths found that over an 11-year period 135, shipboard deaths occurred among German travellers, with a rate of 1.8 per 100,000 and a mean age of 71 for males and 73 for females; the majority of deaths were related to natural causes. Other common problems were associated with hygiene, and as the Centre for Disease Control (CDC) (2020) observed, cruise ships are a crowded, enclosed environment that easily permits person-to-person spread of viruses (e.g. Covid-19 as noted in Chapter 1), and food- or waterborne diseases. Crew members may be persistent carriers of disease according to the CDC (2020). For sea cruising, on-board medical care is crucial, especially given the appeal of this kind of trip to ageing visitors. The CDC (2020) also observed that 50% of all people who seek medical treatment on board are over 65 years of age, with 95% of illnesses managed on board. Typical illnesses included: seasickness (10–25%), respiratory illnesses (19–29%), slips, trips and falls (12–18%), gastrointestinal illnesses (9–10%) and death (0.6 to 9.8 per million passenger nights). The CDC provides clear guidance for GPs on pre-travel advice for passengers who have booked a cruise regarding on-board precautions for remaining safe (e.g. hand washing, food safety, sun protection, avoiding contact with ill passengers and seeking help if ill) (see Hill 2019 and Peake *et al.* 1999 for more detail on cruising, health management and passenger epidemiology). There are also growing concerns about antimicrobial resistance and the ageing traveller among the medical profession, given the overuse of antibiotics and the growing resistance to antibiotic treatments by certain strains of bacteria. The risk increases with international travel and with age. Frost *et al.* (2019) examined the geographic prevalence of antimicrobial resistance, particularly pathogens which commonly affect travellers (e.g. campylobacter, salmonella and other bacteria such as Enterobacteriaceae and staphylococcus, MRSA, VRE (Vancomycin-resistant enterococci) and ESBL (extended spectrum beta-lactamase)) and that are resistant to the common forms of antibiotics passengers might carry with them on holiday. Frost *et al.* (2019) also observed the tendency of travellers on long-haul trips to self-treat with antibiotics. Better pre-travel advice and education to improve infection reduction and personal hygiene is needed, because this is seen as one of the major public health risks over the next 20 years, especially when combined with increasing growth in leisure travel. The transmission of antimicrobial-resistant disease strains globally represents a future challenge for more vulnerable and immunocompromised ageing travellers.

Summary

This chapter concludes the extended discussion we began in Chapter 4 related to the challenge of accommodating ageing visitors who may be affected by a range of medical conditions, disabilities and complex multiple conditions. It is evident that the epidemiology of the ageing leisure and tourism participant is far more complex than we can reflect in this brief discussion. These factors may impose specific limitations and barriers if adequate advice and information to support choices and decision-making are not available or accessed. For the visitor economy, being cognisant of the diversity of experiences that the sector may be catering for with an increasingly ageing demographic means that a sea-change is needed that shifts it away from its current treatment of accessibility, which treats it as a small component of its overall business activity and as a niche area. Instead, we argue that accessibility will need to become normalised and, while specific constraints will always exist where infrastructure cannot be adjusted (e.g. historic buildings), the biggest challenge for businesses is in their own perception and attitudes; it will be necessary for them to make accessibility integral to their offers, not least in their advertising and imagery, so it is not hidden away under an 'accessibility banner' on their website. Likewise, adopting a more inclusive attitude towards ageing is critical to overcoming the identification of ageing as a social problem, which Butler (1969) identified. Since 1969, there has been a concerted campaign to address ageism (Gullette 2017), but societal perceptions and attitudes often take generations to change.

Despite the importance of international and domestic tourism and the growing role which travel medicine needs to play via pre-travel clinics in the future, we should remember that not all leisure has to be undertaken that far away from home. As the evocative title of Bhatti's (2006) analysis of gardens in later life attests, 'When I'm in the garden I can create my own paradise', which conveys the importance of place and space in leisure in that life stage. The pursuit of leisure can take many forms: from in-home to out-of-home, from hobbies and leisure activities through to international tourism. Where the visitor economy is a key component of everyday leisure and less frequent tourism experiences, business awareness and adaptation to create more age-friendly experiences will form part of an innovative shift to meet the needs of a changing demographic. How this might be approached is the focus of Chapter 6.

6   The visitor economy, change and business strategies for ageing visitors

Towards greater accessibility

Introduction: how do we approach the accessibility–ageing nexus?

Ageing populations are complex groupings of people with a right to lead full and rewarding lives, and to enjoy a wide range of the experiences provided by the visitor economy. It is evident that the visitor economies of many countries are not fully prepared for the changes that will need to be implemented as their populations age and the ageing segment of the market for their products and services becomes more significant in the coming decades. The diversity of people categorised as 'ageing' is very broad, and while we might define that category in biological terms, the process is not at all uniform, as the WHO recognises:

> At the biological level, ageing results from the impact of the accumulation of a wide variety of molecular and cellular damage over time. This leads to a gradual decrease in physical and mental capacity, a growing risk of disease, and ultimately, death. But these changes are neither linear nor consistent, and they are only loosely associated with a person's age in years.
>
> (WHO 2018: n.p.)

In this respect, ageist stereotypes are unhelpful and must be challenged so that the contribution of older people and the role that they play are not simply recognised, but developed and stimulated as part of the civil society construct. This requires the barriers posed by 'non-aged' people to be addressed. Sun Life (2019) identified *causal ageism* as one such barrier (i.e. the everyday use of derogatory words or phrases that point to negative connotations about life at over 50 years of age), particularly those associated with social media. Sun Life (2020) also highlighted *retirement ageism*, whereby the over-50s age group is viewed as no longer of value to society due to misrepresentative stereotypes that contribute to invisibility and neglect. As Dixon (2020) suggested, organisations and society frame later life more as something to be endured than as a source of intrinsic enjoyment. Brands sometimes unconsciously reinforce these images, and as Dixon (2020) indicated, unconscious and conscious bias towards ageing people can negatively affect their health and well-being,

DOI: 10.4324/9781003039358-6

and even cause emotional harm in both individual and collective domains. For organisations working in the visitor economy, awareness of these issues is fundamental to building a loyal customer base and in engaging the over-50s age group so as to make products and services accessible, attractive and welcoming to everyone.

This chapter sets out to examine how the visitor economy may need to embrace a more age-friendly approach to leisure in its broadest sense. One key question we focus on in this chapter is – *why do we need change in the visitor economy in relation to ageing, and what frameworks help us to understand and create that change?* Based on our experience and research evidence, the chapter explores the potential for the visitor economy to be transformed through strategic thinking and targeted interventions that embed an age-friendly philosophy. We will demonstrate from both a theoretical and applied perspective the types of action that can promote change. The chapter commences with a discussion of *change* within a business context, given that this is the underlying premise upon which age-friendly transformation is based. We examine how to utilise a knowledge management approach by including *best practice* and *communities of practice* (CoPs) to create the momentum for change in the visitor economy. An example of a CoP is introduced, in relation to dementia, to illustrate the types of tools and techniques which businesses and trade associations/ tourism organisations may use to drive forward change. Such a focus is important because research generated in health shows a disappointing level of incorporation of these tools into professional practice, policy and the public domain. In part, this illustrates the distinct barriers to the adoption of knowledge and to its transfer and translation into practice (Green *et al.* 2009).

The problem of translating research into practice has been described by some researchers as a lack of research pragmatism (Glasgow 2013), while other studies point to a failure to focus on end users' needs. One potential route to overcoming this is the synthesis of new thinking into best practice guides. Such guides often address a specific problem, whereas a CoP has been assembled to look at pooling knowledge and experience in order to translate research thinking and industry practice into guidance for end users. As Wenger and Snyder (2000) indicated, CoPs may help solve organisational problems, transfer best practice and develop skills that enable the organisation to progress, using knowledge as the starting point. CoPs may also be enlarged beyond a single organisation so as to encompass an advocacy and problem-solving role within a specific sector of the visitor economy, and to advance an agenda for that sector. We do not ascribe to the normal academic practice of conceptualising research into a highly complex problem, using vast amounts of intellectual rationalisation and analysis in cases where the issues have already been formulated by a CoP. Instead, we suggest that academia can contribute to helping a CoP on its journey towards the desired outcome of a change in practice and thinking. Otherwise, we run the risk of accepting that academics have all the answers to practical issues, when in fact practitioners working at the interface with people and service delivery have knowledge and experience that can help create valuable solutions. The role of the academic is to support the CoP, and not necessarily to lead it and introduce highly theorised thinking when the CoP is seeking to develop understanding and international knowledge of an issue.

Numerous critiques of business and visitor economy research exist that identify inherent weaknesses in academic contributions that address the big issues of the day; these are reviewed by Phillips *et al.* (2020a) and need not be reiterated here. Suffice to say, translating social science research into practice in an effective and meaningful way is far from an easy process. Ideologically there are many camps and institutional structures and practices that negate this research–praxis approach. In short, business schools (from which a great deal of the visitor economy research emanates), are often focused on the pursuit of esoteric theorisation and publishing research in academic journals with a limited readership; as a result, a great deal of research never sees the light of day and so has no impact on policy or practice. Furthermore, the research is frequently not produced with practical consumption in mind. Applied research is often derided by the academy as lacking intellectual rigour, even if it can offer valuable contributions to problem-solving. As Phillips *et al.* (2020b) illustrated, this has a degree of irony, as business schools are the very places where one would expect research to be put into practice. What we will show in this chapter is that research can bridge this translational gap (see Green *et al.* 2009), in that a CoP can facilitate a pragmatic and practice-oriented focus that absorbs some invited academic contributions. This approach to business-oriented research is rooted in our view that research should make a wider contribution to society in terms of helping to address some of the grand challenges of the day, such as ageing.

Theoretical and conceptual frameworks to create change in the visitor economy

What role for the public sector?

At a philosophical level, there is a broad understanding in many market economies that the state can help to create and promote a better, more accessible and more engaged society, and address civil society objectives. One key focus of this philosophy involves bringing the ageing population into the mainstream of society in an effective manner. To achieve this, greater collective action is beneficial; this should be based on shared interests on the part of NGOs, not-for-profit organisations and communities of interest (as opposed to government) to enable collaboration, communality and participation, fostered without government influence. Where government has a role to play is in setting the public policy framework (e.g. legal obligations associated with accessibility, anti-discrimination and in promoting equality of access for all). While this is not a universally applicable principle adopted by every country worldwide, it is recognised by some governments wishing to progress social development objectives via the United Nations Sustainable Development Goals (SDGs), particularly those in the developing world, as the SDGs focus on many basic human needs (i.e. shelter, clean water and sanitation, zero hunger, reduction of poverty and access to affordable and clean energy). The most relevant SDGs for this book are the pursuit of SDG 3, *Good Health and Well-Being*, to 'Ensure healthy lives and promote well-being for all at all ages' and SDG 10, *Reduced Inequalities*. Applying SDG 16, *Peace, Justice and Strong Institutions*, is vital to ensuring that SDG 3 and 10 are implemented through an inclusive approach to the population and governance. Underpinning such arguments is an orthodoxy associated with welfare economics, a form of normative economics: 'welfare economics focuses on using resources optimally to achieve the maximum well-being for the individuals in society' (Just *et al.* 2005: 3). These arguments have challenged the conventional thinking of positive economics by arguing that through improving the welfare and well-being of the population, economic efficiency and enhanced income distribution can be derived for long-term sustainable benefit. This economic approach can be dated to Pigou's (1920) *Economics of Welfare*, the broad aim of which was to improve the social welfare of the population (see Kumekawa 2017 for a critique).

Furthermore, there is a well-developed literature on public administration (e.g. DeLeon 2008) and the role of the state within it; see for example Karagiannis and King (2019: 2):

> government is necessary to the existence of a civilised society: it delivers basic services, manages the economy in uncertain times, and makes big decisions for a nation's future. The government exercises executive, political and sovereign power through customs, institutions and laws within a nation, and sets and administers public policy. Policy intervention is action taken that 'interferes' with decisions made by individuals, organisations and groups about social and economic matters.

This all-embracing approach to public administration, informed by welfare economics, has been questioned since the 1980s in neoliberal states. The new right, for example, has advocated a minimalist role for the state, enhanced rights of the individual and the creation of markets as a solution to social problems. The new right also advocates a greater promotion of free market capitalism and a shift away from state-led public welfare spending to address inequalities. While economic and political thinkers adopt a variety of different ideological positions on the role of the state, it is recognised that the excesses of the neoliberalism idiom (e.g. Harvey 2007) have created a type of civil society where inequalities have increased amidst a focus on freedom of the individual and policies including deregulation, privatisation and a progressive move away from social provision. Subsequent revisions to these principles have seen a greater recognition that these ideological positions and market-led solutions have for too long been viewed as a panacea for achieving a societal transition, and that ageing can too easily be deemed a problem that marketisation will address (e.g. see Schwiter *et al.* 2018; Feiler *et al.* 2018). There are many critiques of marketisation and the failings in the existing neoliberalism model (e.g. see Vos and Page 2020, who rely on metrics, performance management and market competition), as marketisation has resulted in a focus on control of government spending on long-term social and health care for the ageing population. This has been achieved by promoting public choice and a greater role for the private sector and third sector. This ideological framework has embraced the political discourse on healthy ageing to justify the shift in responsibility from the state to the individual in order to reduce the cost of care that may be avoided by healthier lifestyles. Other approaches advocate a more adaptive strategy towards making provisions for an ageing population, using the language of neoliberalism (e.g. addressing market failure, recognising the market needs of the older consumer, adopting quality of life and well-being approaches, improved information to help with individual decision-making, poverty reduction and addressing age discrimination; see the Joseph Rowntree Foundation 2004).

Through processes of habituation, older people tend to be more accepting of the inequalities and issues faced within society as their experiences become normalised through time. Even so, arguments proposed by Walker (1981: 88) poignantly outline the justification for a political economy perspective towards gerontology:

> A great deal of influential research in social gerontology has tended to treat elderly people as a detached minority, independent from economic and political systems, and 'their problems' in terms of individual adjustment to ageing or retirement. Very little attention has been paid to the structural relationship between the elderly and the rest of society and the differential impact of social and economic institutions on elderly people. An examination of the resources available to elderly people reveals sharp inequalities between them and younger adults, as well as the existence of abject poverty amongst some of them.

These points are also discussed in Chapter 2. Following this strand of research (see Chapter 2 for a discussion on society and structural constraints), Estes *et al.* (1982), argued for a better understanding of the structural constraints that society poses and that cause 'the isolation and alienation of the aged from society, their disengagement, senility, and institutionalization as a function of capitalism and the social relations it produces, and how all of these condition and shape the experience of old age itself' (Estes *et al.* 1982: 161). In the neoliberalist state, with its focus on individual choice, there are two possible approaches to consider in relation to the visitor economy:

- Public sector intervention that may span a range of possible options (see Figure 6.1) and which are more restricted in neoliberalist societies, and

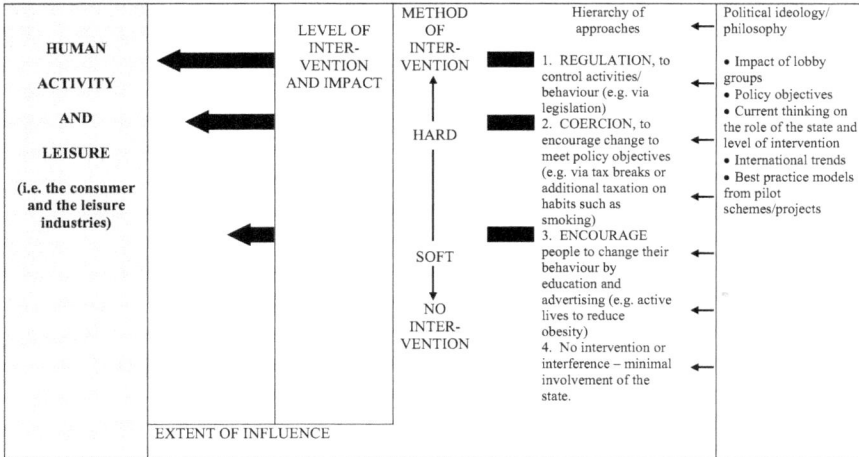

Figure 6.1 Levels of public sector intervention to regulate individual/group behaviour in society: implications for leisure planning and sustainability

Source: Page, S.; Connell, J.; Leisure: An Introduction, 1st Edition, © 2010

Reprinted by permission of Pearson Education Limited

- Adaptation by the private sector/individual business and organisations at a local level, such as work to create age-friendly and dementia-friendly communities, of which visitor economy businesses are a part (see Connell and Page 2019a).

Throughout this chapter, emphasis is directed towards the individual business/organisation, as this is where adjusting operational practices in order to accommodate and ageing agenda may help create change (see Chapter 7 for more discussion of how the WHO has promoted age-friendly communities). For this reason, it is important to have an understanding of how we might progress a change agenda with businesses in the visitor economy, so as to address problems such as ageism, or to make products and services more accessible for specific groups. The following quotation illustrates some of the problems with this latter point; in it, the respondent comments on their experience of organising holidays for ageing people:

> ... we've worked with ... some of the groups who organise pilgrimages to various religious sites ... often their market is the retired adult, you know, who are necessarily thinking in more spiritual terms. So I'm thinking recently of a group of older adults who wanted to go to Greece and they wanted to go around the sites of early Christianity, you know, but again, just the make-up of where these sites are, the physical geography of the places ... internationally ... some of the big chains for hotels are wonderful at having disabled

access … you can go to most countries and find a generic hotel with lifts in it … but where, for example, the provision is more traditional hotels, like staircases or rickety stairs or lots of bedrooms … this has been challenging, you know, the less touristy places have more appeal for older people yet they're so unsuitable [R1].

This illustrates one of the principal challenges to businesses and organisations; the need to match supply and demand so that provision is suitable and can accommodate accessibility needs. At the heart of this challenge is encouraging businesses to rethink how they can adapt and adopt new ideas in order to change their practices and become more age friendly.

How businesses approach new ideas and change: theoretical perspectives

According to Cameron and Green (2004: 2) 'Individual change is at the heart of everything that is achieved in organisations'; so, change is part of the normal evolution of business practices. One area of focus in the change literature is behaviour change (i.e. changing another individual's behaviour in order to achieve a selected outcome such as healthy ageing, via an intervention). Another approach is cognitive change, which, in contrast to behaviour change with its focus on observable behaviour, looks at how an individual's emotions shape how they think. Beliefs and values are important to this, as adopting a positive mental attitude helps individuals go through the different stages of change. The cognitive approach contrasts with the psychodynamic approach, popularised by Kubler-Ross (1969), which examined the psychological states individuals go through during change: death, denial and isolation, anger, bargaining, depression and acceptance. Other stages were outlined by Adams *et al.* (1976): relief; shock and/or surprise; denial; anger; bargaining; depression; acceptance; experimentation; discovery. The Adams *et al.* (1976) study helps to demonstrate that managers need to understand the stages employees pass through when initiating a change management process.

Change management as a dimension of organisational behaviour

Change management for organisations is about reviewing the direction, products and services they offer to customers and how they deliver value, such as profit for shareholders or other organisational missions (e.g. accessibility in public sector organisations). Theorists have described organisational change in various ways, ranging from a chaotic process, a state of constant reflection on the way the organisation operates, through to a more logical and ordered state or almost stage-managed approach. Change management may also focus on events that can create a shock or crisis, such as Covid-19, which may induce change (e.g. taking business online to remain operational). Change management research is a dimension of organisational behaviour (OB), which examines the interactions and relationships between humans (i.e. typically members of the workforce or volunteers) within an organisation at a group and individual level. It also examines how organisations behave (i.e. whether they pursue a change management agenda). Change management research can be traced to the 1960s, when influential studies on innovation (e.g. Rogers 1962) identified the role of people in innovation and change (e.g. early adopters and laggards); subsequent research evolved in the areas of innovation and entrepreneurship. Various models were developed out of diverse areas of social science in the 1970s and 1980s, as the focus on organisational change emerged. Key thinkers such as Handy, Drucker and Mintzberg contributed to the broader dissemination of these developments in organisational theory, which helped to establish the concept of change in the lexicon of management. For example, Handy's (1976) *Understanding Organisations* popularised the study of organisational behaviour, presenting organisational complexity as one way to approach the issue of change. A range of models have been developed to explain the change management process but, as many OB studies illustrate, no model or theory offers sufficient generalisability to be applied to every context.

Key elements of change management models

Lewin's (1947) change model, which is often credited as one of the first in the field (see Hussain *et al.* 2018), was derived from an ethnographic study of change development and comprises a series of stages (unfreezing, changing and refreezing), with a key role for leadership at each stage of change. What the model highlights is change as a process, whereby organisations need to communicate the need for that change in order to bring about a change mentality among employees and to overcome resistance. It focuses on the behaviour modifications required of individuals and their need to adapt in a positive way to enhance their motivation to embrace change. A second model is the McKinsey 7S model with its focus on organisational effectiveness, which identified seven factors within organisations that have to align in order for success to be achieved, as outlined by Waterman *et al.* (1980). The model, which is among the most widely cited in the field, led to the introduction of the 7S framework. Described as a new framework for organisations, it highlighted the multiplicity of factors affecting how a company changes. It emphasised the interconnected nature of the 7S's (structure, staff, strategy, systems, skills, staff style and superordinate goals). The model had no hierarchy or beginning or end point, highlighting the complexity of organisations and the interactions therein. To operationalise this model, the authors argued that a series of steps need to be taken to improve business practices. This model was largely internally focused. A third approach, labelled *nudge theory*, emanated from behavioural economics (i.e. Thaler and Sunstein 2008), and focused on influencing behaviour in order to make a change rather than pursuing a more draconian coercive policy, building upon the issues outlined in Figure 6.1. The principle of nudge theory is to influence cognitive behaviour via a nudge, or gentle push, in order to trigger or invoke a desired change. Various nudge tools exist, such as where a change is offered as the default option, or incentivised by drawing attention to the issue (e.g.

carbon offsetting when flying). Critics of nudging have described it as manipulative and lacking credibility, at least in the short term; on the other hand, Sunstein (2014) highlighted its potential as a form of social engineering. A fourth approach is the *ADKAR five-stage model*. The five stages of this model are:

- **A**wareness of need for change
- **D**esire for supporting and involvement in change
- **K**nowledge of how to achieve change
- **A**bility to make the necessary behaviour change and skill development
- **R**einforcing the change process.

As Prosci suggest (www.prosci.com/methodology/adkar), the model is designed to simplify change management using clear goals and measurable outcomes. It has a people-focused approach and is arguably best implemented in situations where incremental change is the desired outcome.

A fifth approach is Kotter's (1998) *eight-step process for leading change*, which arguably is among the most influential:

1. Create a sense of urgency, of the need for change.
2. Build the change team to lead the process with influential people of all grades.
3. Create a strategic vision of what change is envisaged and the outcomes.
4. Communicate the vision internally.
5. Remove barriers to change, as change management is often a highly politicised activity with vested interests resisting change and innovation.
6. Focus on short-term wins to communicate progress.
7. Maintain momentum to sustain the changes.
8. Institute change and make it part of the corporate culture.

This approach has been found to work well in large organisations, but its weaknesses centre around employee involvement, buy-in and the need for ambassadors and employees as change agents. In smaller organisations, employee engagement may stifle change where emotional response and group resistance may limit support. Kotter and Cohen (2002) analysed 100 organisations, demonstrating that connecting with employees' emotions to spark change was critical (e.g. appealing to employees' and consumers' altruism). External forces along with financial and political factors may thwart change management processes, and so change is an uncertain path to embark on. Kotter's model is very linear and depends upon influential leadership.

In the case of ageing, a great deal of attention has been given to change management by the public sector and charities, as well as individual private-sector businesses, seeking to achieve changes that permeate both their own organisation and society. Models need to fit with organisational ideologies and goals while also appealing to organisations' business acumen and ability to recognise an opportunity (Connell and Page 2019a). When seeking to communicate the broader agenda of ageing or a specific issue related to ageing, one useful approach is a model that has its roots in the WHO's (2012) *Dementia: A Public Health Priority*, which provides a guide to integrating dementia into civil society by setting out the stages that society may pass through en route to becoming dementia-friendly (DF). We have modified this model, as shown earlier in Figure 3.1, to create a conceptual basis upon which to explore whether businesses in the visitor economy fit with the predicted phases of adoption and development of more age-friendly services and products for businesses. The model outlined in Figure 3.1 has been redesigned from the WHO's initial phased approach. In the case of businesses, we have adapted the model to critically evaluate the challenges at each stage of the transition. While any model is a simplification of reality, we suggest that such a framework is important in understanding how businesses can transition towards more age-friendly business processes. In addition, this model reflects the management and innovative capabilities of businesses and their recognition of the benefits of change to their bottom line.

When seeking to understand how businesses might approach the concept of new practices that are more age friendly, it is interesting to note that the current disabilities literature shows that some businesses voluntarily embrace such practices. This may be due to one of a number of factors, such as: the identification of a market segment/opportunity (i.e. economic reasons); for altruistic reasons (e.g. to achieve corporate social responsibility goals; see Segovia-San-Vuan *et al.* 2017); or in response to regulatory measures (e.g. disability legislation). Indeed, some businesses are established precisely with the intention of engaging with what are perhaps defined as 'niche segments', such as people with disabilities, older people and people seeking respite holidays, among others, in the recognition of the growing need to capture new markets for tourism experiences. This literature, combined with the generic business and management literature on the adaptation of innovations (e.g. technology) and the business model innovation literature (e.g. Amit and Zott 2012), suggests that there are distinct stages in how businesses adopt and engage with innovation or new ideas. In the case of the UK, the DF paradigm is very much predicated on a model of innovation diffusion (as discussed below around dementia champions), whereby the spillover benefits of early adopters create a locality or industry sector model so that DF practices permeate into that business area.

Yet adoption of innovative new ideas is not a simple process, as noted in Chapter 3 in terms of Hollenstein's (2004) five major barriers to technology adoption (Figure 3.2), which helps to explain the absorptive capacity or reluctance of individual businesses and their attitudes towards changing business practices. Like technology adaptation, resistance and motivation are primary drivers in adapting this model to DF settings for business. Often the list of barriers is longer than the list of perceived benefits, which may vary according to culture within a business, and the leadership and organisational predisposition to innovation within it. Much of the current thinking around developing DF communities in the UK reflects the epidemic model of diffusion of an idea (see Beltrami 1993). This model is particularly well suited to the voluntary adoption of DF within a prescribed time frame, since the idea was spread initially by dementia champions, ranging from from the Prime Minister's Working Group down to regions and local communities, to enact change at different levels and promote the development of DF communities through Dementia Action Alliances. These champions have tended to be leaders of change who are persuasive, who are risk-takers and who have sought on a limited basis to engage with businesses to sustain innovation in this area. However, the role for businesses centres on voluntary adoption, despite the legal requirements on accessibility of all services (e.g. as enshrined in law the UK by the 2010 Equality Act). As Connell and Page (2019a) found, the relationship between business, societal values and older people with cognitive impairment is encapsulated in the notion of shared value, which helps to engage businesses. In an analysis of local Dementia Action Alliances (Connell and Page 2019a), it was shown that visitor economy businesses and organisations played a role in creating shared value that promoted well-being and quality of life for older people with cognitive impairment. Shared value can be garnered through, for example, a redefinition of products, markets or business productivity, and through different ways of doing business such as collaboration and the development of local clusters (Porter and Kramer 2011) (see Chapters 3 and 7). Before an organisation can decide on whether a change agenda is required and what type of outcome it is looking for, it is likely it will have looked at the role knowledge will play in informing its decision-making process, either based on the knowledge that already exists at management level, or more widely within the organisation, depending upon the size and scale of its operations. For larger organisations, knowledge management is increasingly being recognised as a vehicle to embrace in change management, and so attention now turns to this concept.

Knowledge management

Knowledge management (KM) is how organisations handle knowledge, via different means (e.g. reflective practice, practical or technical means) in order to create value in business relationships and fulfil a purpose by utilising that knowledge productively. Knowledge may be subjective (i.e. it is personal or context dependent) and emerge from experience. Embedded knowledge arises from know-how around existing products and systems. According to Massingham (2020) knowledge is an intangible resource, classified as tacit (implicit) or codified (explicit) knowledge. Tacit knowledge is based in an individual's personal experience and so resides in their mind, whereas codified knowledge is written down or transferred via formal language. Polanyi (1967) found that tacit knowledge is hard to express and share. There is a key debate on whether KM can change organisational culture (Corfield and Paton 2016). Bloice and Burnett (2016) found KM was a major problem in third-sector organisations due to staff turnover, particularly with the use of volunteers (see Chapter 3 on older people as volunteers in the visitor economy), a feature also highlighted by Mahesh and Suresh (2009) and Ragsdell (2016). As McDermott (1999: 104) suggested, KM is an important process for organisations as it can aid in 'connecting people so they can think together'. Yet there are also debates that suggest KM is inherently problematic as an idea, and is even an oxymoron, because 'the more management, the less knowledge to "manage", and the more "knowledge" matters, the less space there is for management to make a difference' (Alvesson and Kärreman 2001: 996). KM is internally focused, and is typically applied to help with innovation and to curate organisational knowledge and opportunities for learning. This means that KM processes may sometimes need to engage a wider range of people within or outside organisations when seeking to develop a broader understanding of issues.

It is in this context that a CoP is often used, whereby a network is formed from individuals who share a common interest or objective. CoP is derived from learning theory, and focuses on best practice and its use in change and innovation. The term CoP was developed by Lave and Wenger (1991) and refined further by Wenger (1998). The community is a forum in which to share and debate ideas and knowledge and so avoid 'reinventing the wheel', thereby improving the change management process. The CoP may also archive knowledge in a written form for reference, storage or to explain best practice, which we will turn to next. The initial focus of CoPs had social participation at its heart, based on the principle of the participant being an active member in the group learning activity. Latterly, CoPs have linked to the area of KM, where know-how and tacit knowledge are formally captured. The pursuit of 'best practice', to which attention now turns, is a widespread objective of CoPs in public, private and third-sector organisations.

Best practice (BP)

In Chapter 3, we briefly introduced the concept of best practice (BP). Its origins as a method of analysis are often attributed to the management theorist Frederick Winslow Taylor (Taylor 1911), who saw it as a means by which to achieve efficiency gains in manufacturing processes, as seen for example in his time-and-motion studies. Yet the first formal use of the term 'best practice' can be traced to the 1960s (Osburn *et al.* 2011). As Osburn *et al.* (2011: 215) argue:

> definitions and descriptions [exist] in more than 50 different fields or areas of endeavour. These are highly variable in their comprehensibility and utility. 'Best practice' definitions and descriptions can be classified in three ways: broad generic definitions meant to apply expansively across the boundaries of fields, disciplines, and countries; field-specific definitions and descriptions (e.g. in the field of disability, or to specialised aspects of such fields, such as physical or intellectual disability); and technique-specific definitions (i.e., circumscribed specifications of 'best practices' in narrow areas of applied practice).

In terms of established broad generic definitions, UNESCO (n.d., cited by Osburn *et al.* 2011) argued that BP can be best summarised as 'successful initiatives … that impact on improving people's quality of life; are the result of effective partnerships between the public, private and civic sectors of society'. In contrast, field-specific definitions of BP tend to focus on the context in which they are developed. In the latter approach, technique-focused definitions emphasise a specific issue or intervention made in order to achieve a change in practice. To summarise, BP refers to a method or approach which peers believe offers superior qualities and has the potential to be adopted as an innovative and/or a standardised way of changing a business practice, method of production or policy direction. Typically, a BP may be made up of a set of guidelines, ideas or practice and may be established and given credence by regulators or organisations that oversee practices (e.g. professional bodies) so it becomes a standardised approach. The benefit of BP in change management is that it helps an organisation to understand what change may feasibly be achieved and where it stands against others (e.g. benchmarking its position). In the public policy arena, it offers a range of alternative approaches to the status quo in order to effect change. BP is a very subjective term; the word 'best' is in the eye of the beholder, and the virtues of any approach need to be verified on the basis of evidence that shows how it has worked elsewhere (e.g. use of metrics) in order for it to qualify for the label 'best'. Bardach (2011) proposed a framework for best practice based on eight stages:

1. Problem definition
2. Evidence gathering
3. Building alternatives
4. Selection of the criteria for best practice use
5. Set out the outcomes
6. Examine the trade-offs
7. Decide which practice(s) to use
8. Tell the story.

Bretschneider *et al.* (2005) proposed an alternative, more technical approach to the evaluation of BPs which examines their methodological and evidence base using a more comprehensive analysis of cases occurs rather than a small sample of examples. BP has had wide-scale use in public health and among NGOs, especially where evidence-based interventions have been proposed in order to make a beneficial difference to people's quality of life. For example, in its Europe 2020 strategy, the European Union (EU) sought to facilitate knowledge sharing of BP more widely. Yet an inherent weakness in this approach is that many organisations are quick to adopt BP without a detailed evaluation of its fit to their situation or of whether the organisation's current direction of travel is adequate to enable it to adopt the changes needed. There is a tendency to believe existing practices are inadequate, and thus, new and better practices are needed. However, the real value of BP lies in its potential as a means of approaching a new problem or opportunity that the organisation has not addressed before. There is also an acceptance that we need to transition from BP once it has been established and adopted so that it becomes *good practice*. In theoretical terms, BP has been linked to innovation management and organisational success. A focus on five key factors is important to the successful implementation of BP. These factors are: strategy and leadership; culture and climate; planning and selection; structure and performance; and communication and collaboration. Numerous models BP implementation also point to the importance of project management, staffing, resources, timing and outcomes, developing a workable action plan, and access to a selection of a range of BPs to address the problem. BP needs to avoid becoming an end in itself and distorting the corporate mission. Ideally BP is an innovation, applied to solve a specific problem and to give the organisation a competitive advantage (Porter 1985), helping it to differentiate itself from its competitors.

An alternative approach to BP proposed by Hansen and Birkinshaw (2007) views innovation as a value chain, a means of transforming ideas into practice rather than just importing BP. BP may help to sharpen up the implementation of change processes but is not the sole solution as an organisational champion is needed to promote change. As Dembowski (2013) suggested, enablement and reducing barriers in order to increase capability or opportunity are crucial to achieving the aims associated with BP. Yet Kanter (2006) indicated that relying upon the innovation/BP of other companies may be problematic, as it does not harness internal creativity.

Obstacles to BP: resistance to change

There is a well-developed literature on the obstacles that may arise when organisations seek to promote a change agenda; this can be traced to Loch and French (1948). Dent and Goldberg (1999) define the Resistance to Change (RtC) concept as describing a situation where 'employees [who] are not whole heartedly embracing a change that management wants to implement' (Dent and Goldberg 1999: 26). One perspective within the RtC literature is to create 'change agents' to overcome it, although these in turn may also create resistance. In terms of the sources of resistance (e.g. myopia, denial or refusal to accept information presented, misinformation or misunderstanding, organisational silence, the speed of change, reactive mindsets or poor strategic vision), these may reflect deep-seated interests and political positions, imbued with cynicism and poor consideration of the social impact of change on people (Del Val and Fuentes 2003).

Engaging the visitor economy to facilitate change: tools and techniques

Connell *et al.* (2017) demonstrated the importance of collaboration within the visitor economy among professional and trade associations to achieving a unified form of communication able to reach diverse types of organisation, ranging from private-sector SMEs, through public-sector-funded bodies like museums and charitable funded attractions, to international corporations with multiple operations. Some successful models of engagement have been documented in the tourism sector (e.g. Franchetti and Page's 2008 study of the effect of CoPs on innovation). But the sector's fragmented, unconnected and often insular approaches remain a persistent weakness that affects how the visitor economy engages with (a) other businesses within its sector; (b) other businesses in other cognate sectors; (c) with government

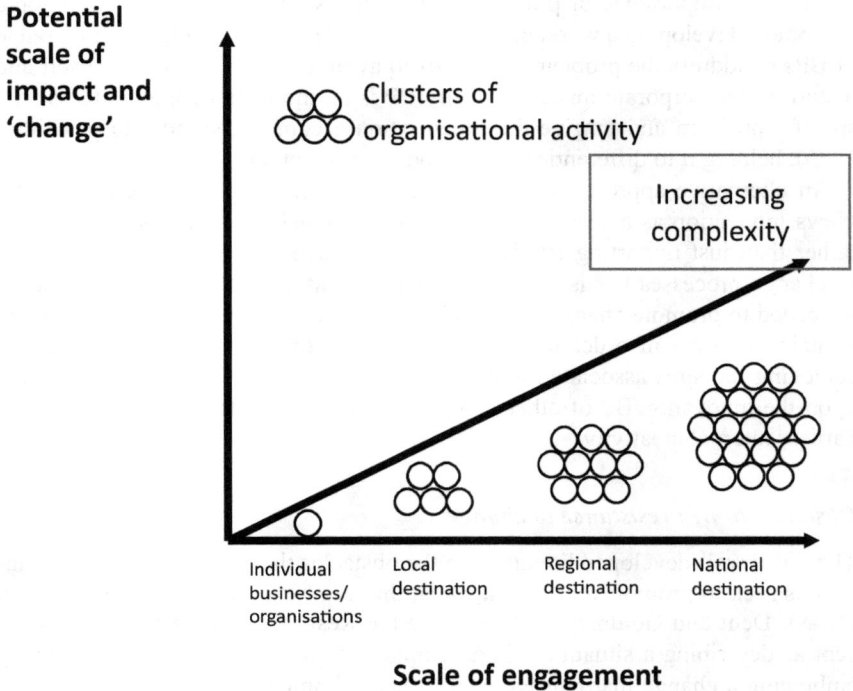

Potential scale of impact and 'change'

Clusters of organisational activity

Increasing complexity

Individual businesses/ organisations

Local destination

Regional destination

National destination

Scale of engagement

Figure 6.2 The engagement continuum in the visitor economy

and policymakers, typically via trade associations; (d) the media and public; and (e) international bodies such as the United Nations World Tourism Organization (UNWTO). In many countries, bringing the visitor economy into a more loosely organised amalgam of interest groups, typically represented by the professional and trade organisations, is the role of public- or private-sector-funded bodies such as national, regional or local tourism organisations (Connell and Page 2019a).

Given these issues of engagement (which, as discussed earlier, are compounded by poor uptake on the part of the university research sector), we can discern the four critical points of engagement for the dissemination and implementation not only of BP, but also of innovative solutions in terms of accommodating and nurturing ageing consumers (Figure 6.2). Figure 6.2 illustrates a continuum of engagement ranging from individual business to multiple organisations at a national scale (or beyond). A notable feature is that engagement may occur simultaneously across the entire continuum, adding two distinct layers of complexity. First, the issue of scale means it is potentially easier to interact and engage with one organisation or business than with many. Then, as we move through the continuum, engagement becomes progressively more complex, with the involvement of multiple stakeholders and individuals. This is why the visitor economy tends to use industry-led trade associations to speak collectively with one voice, an effort that is complicated by the involvement of destination management organisations that are also trying to coordinate, facilitate and engage interest on the part of these associations and individual businesses in parallel (Connell and Page 2019a). A second, and perhaps more problematic issue is that as one moves away from the individual organisation, achieving a consensus becomes more challenging due to vested interests and the politics of business (Page 2019). Not only does this create complexity, but it potentially weakens the perception and image of visitor economy in terms of its ability to engage with a change agenda, as Page (2019) outlined in terms of the snowball effect of tourism development and the problem of vested interests not wishing to see tourism curbed or changed. With much of the change that will be required for adapting to ageing consumers, and the growth agenda of the visitor economy globally, identifying a new market (e.g. ageing) would appear to be the main hook with which to address the change/adaptation agenda, as Connell and Page (2019b) found.

However, other models of engagement can be of equal value in promoting a change agenda: namely the CoP model, to which we now turn.

The experience of dementia and the visitor economy: measuring and evaluating success as a blueprint

Implementing change to overcome complacency, the status quo and RtC, where it exists, remains a challenge. While effective communication is vital in transformative change, building a team and a shared culture is one important tool that many organisations have adopted by creating *champions*. The roll of a champion is to act as an advocate to achieve employee engagement by demonstrating their passion about the change agenda, often as a way of providing feedback to senior managers and of connecting with people in diverse ways. Champions may be a conduit for innovation and change, as Coakes and Smith (2007) show, through their multiple contacts and the communities in which they participate. In the case of dementia, its link with the visitor economy had been overlooked prior to the studies undertaken by Page *et al.* (2015), Innes *et al.* (2015) and Connell *et al.* (2017; Connell and Page 2019a, b). To champion change in business practices in response to dementia, various forms of university-to-industry dissemination have occurred, beginning with the pioneering study by Page *et al.* (2015). Based on the work of Connell *et al.* (2017), an initial output was an industry guide for the National Coastal Tourism Academy in 2015 (https://coastaltourismacademy.co.uk/resource-hub/resource/dementia-why-is-it-important-for-tourism), which was designed to increase awareness of customers with dementia among visitor economy businesses. An industry summary report for the Scottish visitor-attraction sector, based on a study of awareness and actions taken at attractions was produced to feedback summary results to over 500 tourism businesses, assisted by a trade body – the Association of Scottish Visitor Attractions (www.asva.co.uk). These initial dissemination exercises sought to develop a greater understanding of how to make the visitor economy DF (mainly focused on Stage 2 of the model outlined in Figure 3.1: 'Some awareness, possibly from personal experience or awareness-raising activity'), and were intended primarily as an awareness-raising tool.

As Connell *et al.* (2017) found, visitor economy business engagement with dementia showed some or no awareness of the issue, the exception being the leading advocates (i.e. where champions existed) in the museums sector. Several barriers to engagement were identified, including perceived costs of adaptations, and lack of time or expertise. Where DF initiatives did exist, they were found to be sporadic and non-uniform, suggesting that advice and resources are needed to help the visitor economy to grow its DF infrastructure, in order to progress towards Stage 3 of the model outlined in Figure 3.1. Specifically, knowledge, resources, organisation and support were needed to build the DF infrastructure so that it could move beyond the 'awareness' stage. What Connell *et al.* (2017) demonstrated was that businesses wanted to engage but did not always have the 'know-how' to do so, which illustrates the critical role of BP, CoPs and destination management organisations (DMOs) in helping with such tasks. Connell *et al.* (2017) concluded that clear, tailored and targeted information on practical strategies was needed in order for businesses to adopt DF approaches. This was reinforced in Connell *et al.* (2019a), in which the activities of DMOs (see Chapter 3) were evaluated in order to assess what progress and actions had been taken by the tourism sector to manage this process.

To fill these knowledge and know-how gaps, and focusing on practical approaches, two key initiatives using a CoP methodology were developed by industry champions in 2017 and 2019, in which the authors were partners, building upon their research on DF practices. The first was the Historic Royal Palaces (HRP) (Historic Royal Palaces 2017) *Rethinking Heritage Guide*, developed as a collaborative outcome from a CoP formed by HRP in 2016 as an output from the DF Heritage Network, also convened by HRP. The second was with the national tourism organisation for England, VisitEngland, in 2019.

The 2017 guide was predominantly drawn up and co-authored by heritage practitioners. This practical guide sought to draw together current research and best practice advice from the visitor economy to enable the heritage sector to move the issue forward (which we describe as moving DF approaches beyond Stage 2 of the model in Figure 3.1). The authors were recommended by the Alzheimer's Society (AS) to be co-opted to the CoP, based on their previous research studies on the DF visitor economy and their contribution to the Prime Minister's Aviation Working Group on Dementia.

The resulting guide (www.hrp.org.uk/media/1544/2017-11-14_rethinkingheritage_lowres_final.pdf) was a 60-page document targeted at managers in the heritage sector, with 610 hard copies produced in 2017 accompanied by 250 PDF downloads from the HRP website and a further 250 downloads from the AS website by February 2018. The purpose of the guide was to help heritage site managers and volunteers to:

- Gain a better awareness and understanding of dementia
- Build a business case for DF heritage practice (i.e. to have an advocacy role)
- Learn top tips through practical examples and case studies
- Feel inspired to make their site DF.

A subsequent evaluation of the guide and its impact upon professional practice a year after publication concluded that:

- The guide increased awareness of DF practices among those already aware of the issue. Those who were unaware of the issue prior to reading the guide became more informed about it. Comments from a number of respondents about the value of the guide stated that it was valuable in achieving DF goals, as the following quotations suggest:

    1. 'Heritage is very definitely in the health and well-being area now and therefore social inclusion of all sorts should be the over-arching priorities of an organisation.'
    2. 'I think they [people with dementia] should have equal access to the site – as long as the provision is in place, they should be able to have a chance to engage as effectively as other visitors do.'
    3. 'Heritage does have a role to play and I think it's about choice and ensuring that people living with dementia have the range of choices that any of us expect.'

These quotations illustrate the importance of achieving inclusivity, equity, access and provision for those people with dementia and their carers who wish to consume heritage experiences.

- Guide users were also able to evidence changes and adaptations to their practices: one respondent felt it was *'a good advocacy tool'* and that *'audit examples in the guide [were] particularly helpful'* with a *'good balance of case studies and examples';* they described it as a *'flexible high-quality resource but not an end in itself. I think it is part of a process of how we think about choices and opportunities in terms of a broader equalities agenda'.* In particular, *'case studies were really, really useful because they give practical advice and also advice on how to get the rest of your organisation on-board with the new initiative's result of increased awareness after reading the guide'.* One respondent concluded that *'I think just on a really simple fundamental level it's positive, it's creative, it's the art of the possible which, you know, is what the heritage sector needs.'*
- A wide range of new activities and changes to existing practice were implemented after users had read the guide (e. g. health walks and guided walking tours around the site, specifically adapted to be DF; learning/craft/reminiscence sessions; memory boxes to stimulate reminiscence and nostalgia; taking talks and slideshows on local history /heritage into care homes; including sensory and stimulating experiences in gardens and exhibitions).
- Where users of the guide identified remaining barriers to becoming more DF, such obstacles were primarily related to the structural nature of the specific heritage site or the sustainability of funding for specific DF initiatives.

- The guide also had a wider range of uses, as its users found that they were able to incrementally build further behaviour and practice changes to create DF environments through training opportunities for users in their organisations. This was because the guide had disseminated various innovations in practice and included examples of BP in the case studies.
- The user orientation of the guide made its content and knowledge transferrable, which achieved the aim of users being able to 'Learn top tips through practical examples and case studies'. Respondents mentioned that it provided a *'good balance of case studies and examples'* and that it was a *'flexible high-quality resource but not an end in itself. I think it is part of a process of how we think about choices and opportunities in terms of a broader equalities agenda'.* Case studies were considered *'' really, really useful because they give practical advice and also advice on how to get the rest of your organisation on-board with the new initiative'.* One of the key features that helped managers was *'reference to the dementia pound and the idea of the business case and I think that's really useful because I think we need to get away from an idea that ... the heritage sector ... [and our] offer for people living with dementia is a kind of nice-to-do or a charitable act, you know; there are [sic] business case for thinking about inclusion.'*
- The timing of changes and adjustments within organisations was identified by the guide users as between 6 months and 2 years.
- Further developments of the guide were identified that involved minor changes to the content and a general updating in three strategic areas: continuation of the advocacy aims embodied in the guide and a relaunch following an update, after which dissemination of the guide continued.

The guide's wider impact upon heritage policy was noted by the Heritage Lottery Fund (HLF) in their inclusion guidance, launched in January 2019, for organisations seeking funding from the HLF to enhance visitor provision and site enhancements. In late 2019, England's National Trust (NT), with 500 heritage properties, announced it was creating a three-year plan to make its sites DF, in collaboration with AS, one of the partners in the guide's development. The NT receives over 26 million visits a year, and of its 5 million members, it estimates that around 150,000 may be living with dementia. Justifying the move towards becoming DF, the NT cited research from the AS that was able to demonstrate the importance of such sites for people living with dementia as venues for day trips; this was because these heritage spaces were safe and familiar spaces for this group. The NT cited Innes *et al.* (2015) as evidence of their claim.

A subsequent refinement of the guide principle, targeted at a much broader visitor economy audience, was developed in 2019 based on the CoP methodology coordinated by VisitEngland's Inclusive Tourism Action Group (EITAG). EITAG is a policy-focused working group representing the national tourism sector. The CoP comprised staff from VisitEngland, the NT, VisitScotland, AS and the authors of this book, with advice and case studies sourced from businesses that were innovating in this area. In contrast to the HRP (2017) guide, the VisitEngland (2019) guide was much shorter: (30 pages) in A5 format. Its underpinning principle was simplicity, with a focus on the steps required in order to become DF, as illustrated by the guide's contents (*Introduction; Why become dementia-friendly?; What is dementia?; Living well with dementia; Information; People; Place; What can you do next?*). A VisitScotland version of the guide was also published simultaneously under the Advice Link brand. The VisitEngland (2019) guide followed a similar model to other guides the organisation had produced in partnership with stakeholders on autism and deafness to address inclusivity issues in tourism. By early 2020 the VisitEngland guide had been embraced by 31 organisations within their own organisational work, but it had been disseminated to a much wider global audience in countries such as China and across Asia, illustrating international reach in terms of sharing BP.

Can these best practice guides be used more widely in the visitor economy?

The two CoPs discussed here, as well as their outputs, mark an important step forward in terms of visitor economy research, bridging the gap that previous initiatives have tried to address between academic research and industry practice. Previous examples of attempts by academic bodies to produce summary research reports for industry have had limited success in making a significant impact on industry practice, largely due to the gap between industry and academia. These two guides drew upon the KM model of a CoP to derive BP examples, using current research data on dementia and the visitor economy. One of the continued challenges that academic-generated reports for industry face is their failure to embrace the needs of managers. Lengthy or academically robust research justifications of methodology and theory are not needed for industry-facing publications. That is taken for granted when industry bodies seek academic advice and input. The principal barrier is the failure to get advice or guidance on what an industry reader requires. Both BP guides were designed and developed in a glossy, illustrated, engaging and enjoyable-to-read format, with key ideas, checklists, top tips and case studies to inspire thinking and stimulate action. This is the polar opposite of academic journal articles, which use a style that tends to communicate to a narrow readership of other academics. The two CoPs reported on here built on a set model used by the AS and VisitEngland for their BP to create a tool that works in communicating with its target audiences of business owners and managers. The other element that is often underestimated is the critical role of the CoP champion in leading the guide-creation process, alongside a practitioner who understands how guidelines will be adopted and used. In addition, the model in Figure 3.1 maps out different stages in the journey towards making a DF experience into a normalised experience. Our understanding is that individual organisations start at different stages of the model when they engage in the DF journey. So 'change' is a gradual process, typically conforming to models of innovation diffusion and adoption whereby champions in organisations help to push the innovation process while other organisations observe the changes and catch up, and others prefer to ignore the problem.

To return to the question – *can this best practice model be used more widely in the visitor economy?* – it is clear that BP models offer great traction for addressing specific issues associated with ageing (e.g. dementia) and more generically with ageing. The weakness is that BP tends to rely upon highly motivated individuals to champion the issue, who then need to convene a CoP when creating guidance: this requires good networking, organising and project management skills along with effective management of meetings. We also need to recognise that the co-creation role of the CoP in creating guidance can be very time-consuming. Nevertheless, creating guidelines alone should not be the end of the process; it should be the start, followed by evaluation of their impact and tracking of 'change' in the visitor economy, typically over a short time frame (i.e. up to three years), as was the case of the HRP guide. At that point, published guidance may need a refresh and relaunch to give it a new lease of life and to reach new audiences.

Summary

This chapter has illustrated the importance of the tools which can be used to bring about a more age-friendly visitor economy, and how BP and CoPs can help create a momentum for change when deployed creatively. The case of dementia illustrated the types of tools and techniques that can be created by a CoP and that businesses and trade associations/ tourism organisations may use to drive forward change beyond the confines of academia. Where that CoP model is actively developed, it overcomes the translational problem of taking research and putting it into practice by focusing on end users' needs. We explore this theme further in the next chapter as the CoP model provides a workable solution to the age-old problem of creating academia-industry-policy interactions and ensuring that academic knowledge is developed to create pragmatic solutions. Yet we should not underestimate the problems of organisational readiness to implement change (Weiner 2009) and the continuing problems that translating research into practice pose (Nilsen 2015).

7 Future agendas

Making the visitor economy more age friendly

Introduction

This book has emphasised that ageing is an issue for the global visitor economy as population structures in many countries across the world skew towards older age groups, and the proportion of older consumers and leisure participants expands. As Hyde and Higgs (2016) recognised, ageing is a universal human phenomenon and the experience of ageing manifests itself at a range of spatial scales from the global through the national, regional and local levels (e.g. the neighbourhood), right down to the microscale of the home, the family and individual. The real challenge, as this book has hopefully demonstrated, is what Carney and Nash (2020) refer to as *a lack of public understanding of ageing*, especially within the visitor economy, given the complex issues which arise from an ageing demographic. Ageing has important implications for families, communities, businesses and society, both now and for future generations. A third of children born in the UK in 2021 may be expected to live to 100 (Economic and Social Research Council 2021). There are also numerous predictions that we will enjoy healthier, longer lives as part of that increase in life expectancy. Many assessments of ageing in the future point to how the individual's experience will be shaped by their life course, although currently much of our knowledge is derived from a North American/Euro-centric perspective. This overlooks the changes occurring in the Asia-Pacific, Africa and Latin America. Although the number of non-English-language publications on ageing is expanding, the changes occurring as a result of rapid urbanisation in the developing world are creating future localities where ageing will need to be addressed. Not surprisingly, the application of age-friendly criteria to the developing world has been limited (Rudnicka *et al.* 2020) although at a global scale, ageing is starting to impact the visitor economy, not least – as we have shown – in terms of the work of tour operators, the hospitality sector and destinations. This chapter is a synthesis of the thinking developed throughout the book and focuses on one key question: *how do we make the visitor economy more age friendly?* The chapter commences with a discussion of the predicted global drivers of social and economic change through to 2030 that will continue to shape the ageing population and influence ageing-consumer trends in the leisure and tourism domain. This is followed by an examination of the parameters of an

DOI: 10.4324/9781003039358-7

age-friendly environment, what this means for leisure and tourism environments and the types of tools needed to implement age-friendly principles in the visitor economy. One key challenge is in seeking how to envision the visitor economy as it adapts to ageing in the near and distant future, and we turn to this issue first.

The visitor economy in 2030 and 2050

Ageing follows a series of stages through an individual's life, regardless of their calendar age, which progress through independence, interdependence, dependency, crisis management and end of life. Each of these stages has meaning and value for the individual, their families and friends, and as we have shown throughout the book, these stages interact with the visitor economy. Even death has a celebratory aspect in many cultures, ranging from private and personal events with families and friends to larger-scale public celebrations of a person's life. These celebrations are facilitated by the visitor economy, as we highlighted in Chapter 5, even though marking the passing of a loved one is a sensitive subject and one that remains somewhat taboo in studies of the visitor economy.

Conceptualising the ageing traveller is important as ageing frequently determines the extent to which people may experience a greater range of health issues or multiple conditions. For leisure and tourism decision-making, stage of ageing and presence of health considerations will create a greater demand for pre-travel advice on the suitability of certain forms of travel and activities for ageing travellers, the precautionary measures needed (e.g. inoculations) and predisposition towards certain environmental constraints on travel. For the supply of tourism experiences and infrastructure, this will mean that more age-focused provision that recognises visitor needs will be required. From a leisure perspective, by 2030 we are likely to see a rise in both multigenerational families and single households, many of which will be more familiar with and embracing of technology than in 2021. For ageing travellers, a greater use of artificial intelligence (AI) will be rolled out for many of the touchpoints visitors engage with when planning, booking, travelling to and spending time at destinations, events, accommodation, attractions and airports. But Holley-Moore and Creighton (2015) indicate that mobility of the ageing population revolves around driving, which is their most common form of transport in many developed countries for older people. Alternatives like public transport remain poorly planned and developed for ageing travellers outside of urban centres. To facilitate simple day-to-day interactions with transport-related infrastructure (e.g. pedestrian crossings), some of this infrastructure may need to be redesigned, as for example the time allowed to cross a road is too short on automated crossings for some ageing people. The mobility of an increasingly aged population may be accompanied by a greater use of autonomous vehicles, non-motorised vehicles and transit options (i.e. assistive technology) that are yet to be invented and rolled out. It is not unreasonable to suggest that various forms of leisure-related travel will be linked to personal carbon budgets as oil becomes a thing of the past for all forms of human mobility. The decarbonisation of the visitor economy and new hydrogen- and electric-based technologies will power the highly consumptive tourism and hospitality sector and may lead to more widespread adoption of the user-pays philosophy. This shift in philosophy will be informed by the continued loss of valued environments (e.g. the Arctic and Antarctic ice sheets), and loss of biodiversity. We may finally have to admit that tourism no longer offers the sustainable development options that the sector has used as a rationale for promoting tourism-led economic development since the 1980s. The sector has deliberately used greenwash techniques and a smokescreen of sustainability to make travellers feel better about the impact of their impacts on the environment. The ageing tourist in 2030, and certainly by 2050, will be more knowledgeable about climate change than previous generations, and more affected by the changes in the global climate that will increasingly influence leisure and tourism behaviour.

Sustainability as applied to tourism is highly contested as a misused ideology that has failed to stop the excesses of human consumption related to mass leisure and tourism. A greater focus on localism and appreciating the diversity and variety of one's own environment and country may reinvigorate the domestic tourism sector and weekly leisure opportunities for an ageing population. Virtual reality and new technologies will help a generation that has had unrestricted travel to reminisce and reconnect with places visited and experiences enjoyed. Hotspots of concentrated leisure and tourism activity in developed countries will most likely lead to more regimentation and pre-planning of consumption opportunities that may suit the ageing visitor, whereby they are able to travel off-peak and still have flexibility. Yet this poses ethical issues about whether developing countries will seek to sweat the environmental assets for tourism, positioning themselves as premier products to capture much-needed foreign-exchange revenue, or whether foreign aid might help divert their economies away from the current overdependence upon tourism expenditure. As countries invest in programmes of healthy ageing to reduce the burden on health services and target the twin problems of obesity/inactivity and social isolation/loneliness, this may intensify the demand for tourism and leisure services.

Understanding future tourism and leisure trends

The UK Cabinet Office Performance and Innovation Unit (2001) explored ways of thinking about the future in terms of alternative possibilities that can be grouped as: '*what may happen* [possible futures], *what is the most likely to happen* [probable futures] *and what would we prefer to happen* [preferable futures]' (Page *et al.* 2010: 101–2). One of the most commonly used techniques is forecasting using quantitative data (Figure 7.1) spanning a continuum of soft to hard tools, drawn from diverse disciplines and management science (see Page *et al.* 2010). Amadeus's (2015) *Future Traveller Tribes 2030* considered the question: how will purchasing habits evolve? The study identified six types of future traveller tribes:

- *Obligation meeters*, who travel to meet a specific obligation (e.g. business travel, family event or routinised trip);

SOFT TECHNIQUES   OBJECTIVE OF TOOL

Informed fiction writing   Intuitive / Learning Embraces

Behavioural simulation

Scenarios ← More qualitative than quantitative
           ← Recognises uncertainty
             Clarifies risk

Modelling

Forecasting ← Focuses on certainty
            ← Disguises uncertainty and conceals risks
              Single-point projections

Extrapolation

                                                         Analytical / Control
HARD TECHNIQUES   Aiming for Certainty

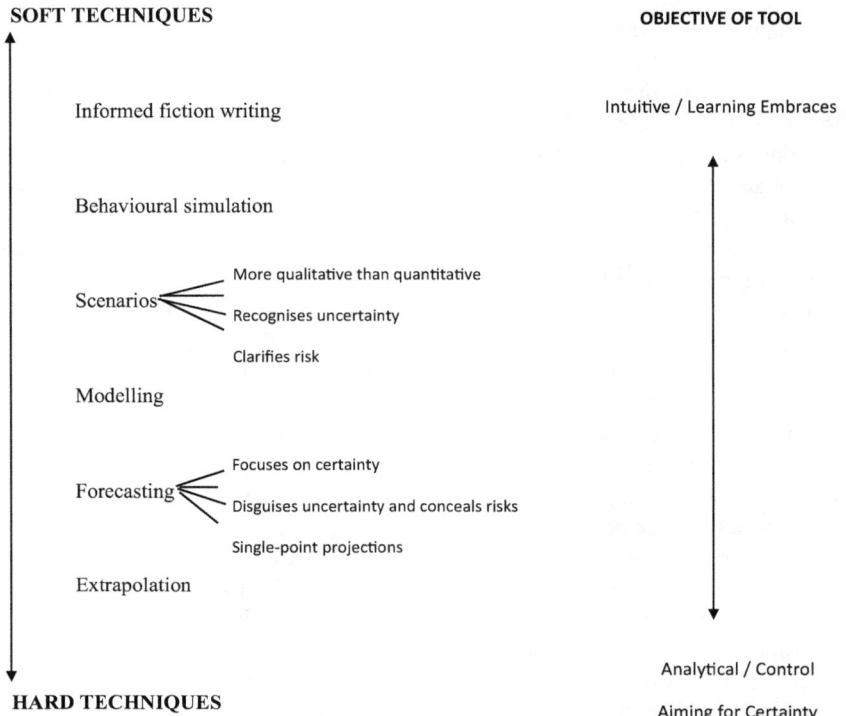

Figure 7.1 The practices used in future studies

- *Simplicity searchers*, who seek simple solutions without having to undertake extensive searching of options;
- *Ethical travellers*, whose conscience and values shape their travel (e.g. environmental focus);
- *Reward hunters*, who focus on indulgent and luxurious travel, for whom the reward reflects their hard work and so travel is a motivational tool;
- *Cultural pursuits*; for this group, the trip is an opportunity for cultural immersion and an experience that is different from their own culture;
- *Social capital seekers*; this category is a well-travelled group seeking the social reward for their trip via being watched by an online audience.

Technology may provide an opportunity for a more tailored, personalised experience (not dissimilar to what Poon (1993) described as 'new' tourism) in that it has digitised the different stages of the trip cycle (e.g. *inspire – shop – book – post – booking – pre-departure – journey to airport – airport experience – the trip – post trip*). What has changed radically since Poon's analysis is the rise of smart technology enabling micro-personalisation of the trip cycle, with each stage becoming a perpetual touchpoint via digital connectivity. By 2030, many, if not most, ageing consumers will be digitally savvy and so able to embrace advances in smart technology, which will permit even greater tailoring of the experience. As Respondent 6 indicated in terms of future trends, *'Every year, more and more of our participants are using our website and finding out about us through digital channels'*, confirms the growing importance of these channels. Amadeus (2015) identified ways in which the visitor economy may harness technology in future to interact with ageing consumers (e.g. using advanced robots to deal with routine services processes like check-in; future artificial intelligence enhancements will make it possible for the robots to respond to consumers' emotional needs). Specific obstacles also exist in terms of ageing visitors' mobility, especially they need to use when public transport. Cirella *et al.* (2019) highlighted possible innovations for ageing passengers around accessibility, affordability, availability and acceptability. Flexible transport services (e.g. Dial-a-Taxi) with adaptations do work well for the mobility-impaired, although Holley-Moore and Creighton (2015) found that ageing people gave up driving only as a last resort. Cirella *et al.* (2019: 15) cautioned against optimism, as:

> On a global scale, it is evident that transport innovation for elderly people is country-specific with distinctive decision-making, policy and enforcement regulation. The majority of the literature is focalised [sic] in the USA, Canada, Australia, New Zealand, Japan, Hong Kong and Western Europe .... A huge gap in the research is information from all low-income countries. These perspectives and potential innovative, and perhaps low-cost, solutions are missing.

Luiu *et al.* (2018) described the unmet travel needs that can transpire in later life; these range from *primary* (or utilitarian) needs to perform necessary tasks like shopping to *secondary needs* (e.g. those that meet affective needs such as maintaining independence) and *tertiary* travel needs (such as for recreation and tourism). From an ageing (and disability perspective), Helal *et al.* (2008) examined how smart technology could help address unmet travel needs as a major enabler. Loos *et al.* (2016) describe smart technology as part of a *mobile digital ecosystem* with its capability to link visitor supply and demand so as to create age-friendly visitor environments. The new technology around smart systems not only improves accessibility through improved real-time information provision, but in identifying opportunities for increased mobility to reduce loneliness and isolation within the ageing population. Battarra *et al.* (2019) illustrated how smart technology helps us to recognise the critical links between transport needs, land use, time, individual needs and visitor environments. But as Cirella *et al.* (2019) highlighted, this has been rolled out in an ad hoc way in the form of trials or initiatives to improve public transport and enhance transport networks, or in softer mobility settings (e.g. people travelling on foot) and through platforms developed to help with travel planning (e.g. Traveline in the UK). Helal *et al.* (2008) also highlighted how smart technology could be engineered using assistive devices and systems for people with visual, mobility, hearing and cognitive impairment to enhance their lives, in ways that include leisure uses. Other developments are expected in human–machine interaction and virtual reality, and in the use of assistive robotics which will continue to transform our lived experience of tourism and leisure. Smart technology may help to address cognitive and dexterity issues that are compounded by age.

Age-friendly research for the European Union (EU) highlights that the key barriers for an ageing population include wheelchair access, medicine availability, reforming the single-room supplements for single older travellers and transport provision in destinations where public transport is not fit for purpose. Awareness of these issues is minimal. Much of the research has been directed at policy-focused outcomes (e.g. age-platform.eu). For example, in 2018, Age UK Bristol sponsored an *Age-Friendly Tourism Business Award*, while Silver Group Asia (www.silvergroup.asia) examined the age-friendly nature of tourism in Singapore, identifying 350 touchpoints at the destination. This illustrates the importance and burgeoning awareness of age-friendly issues for the visitor economy.

Creating an age-friendly environment for leisure and tourism

Much of the impetus for developing age-friendly communities is derived from the WHO Age-Friendly Community model (WHO 2007b; Figure 7.2), which is designed 'to improve … physical and social environments to become better places in which to grow old' and the creation of the WHO Global Network, which consists of 830 cities and communities in 41 countries that are working towards becoming age friendly (see Table 7.1 for the localities in the UK).

More specifically, the WHO also set out its methods for progressing the age-friendly community agenda:

> Creating environments that are truly age-friendly requires action in many sectors: health, long-term care, transport, housing, labour, social protection, information and communication, and by many actors – government, service providers, civil society, older people and their organizations, families and friends. It also requires action at multiple levels of government. The following key approaches are relevant to all stakeholders:
>
> - Combat ageism;
> - Enable autonomy;
> - Support Healthy Ageing in all policies at all levels.
>
> WHO raises awareness on the importance of environments in determining Healthy Ageing and encourages the creation of age-friendly environments by:
>
> list
>
> - Compiling evidence-based guidance on age-friendly environments;
> - Providing an information platform for sharing of information and experience; and

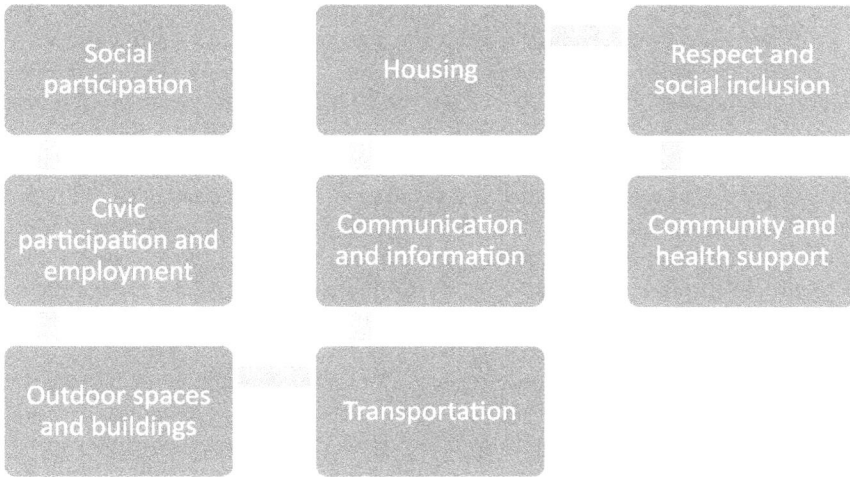

Figure 7.2 Elements of an age-friendly city

Table 7.1 Age-friendly communities in the UK

Antrim and Newtownabbey, Northern Ireland	Glasgow	Middlesbrough
	Greater Manchester	Newcastle Upon Tyne
Ards and North Down, Northern Ireland	Hastings	Newry, Mourne and Down
Banbury, England	Isle of Wight	North Yorkshire
Belfast, Northern Ireland	Leeds	Nottingham
Birmingham	Lisburn and Castlereagh	Salford
Bolton	Liverpool	Sefton
Brighton and Hove	Liverpool City Region	Sheffield
Bristol	London	Stockport
Cheshire West	London Borough of Lewisham	Stoke-on-Trent
Derry City and Strabane	London Borough of Southwark	Torbay
East Lindsey	London Borough of Sutton	Trafford
	Manchester	Wigan
Fermanagh and Omagh	Melkham	York

Source: Developed from the Centre for Ageing Better, www.ageing-better.org.uk

- Nurturing and developing the WHO Global Network for Age-friendly Cities and Communities (www.int/ageing/age-friendly-world/en)

A number of localities have embraced the WHO Age-Friendly Community model (e.g. Andalusia, Brazil and Australia), while Canada has been a leader in the creation of age-friendly locales (Plouffe and Kalache 2011). Canada's Ministry of Health provided considerable encouragements and guidance for communities on how to navigate the process of turning an ambition into reality. The WHO criteria for becoming an age-friendly community recommends that communities follow the steps of their *Age-Friendly Cycle*. This involves: sending a commitment letter to the WHO Age-Friendly Network to register their intent; creating a baseline assessment of age-friendly infrastructure needs; devising a strategy and action plan; implementation; and evaluation. The Government of Canada's 'How to Guides', such as its *Age-Friendly Communities Implementation Guide* (Public Health Agency of Canada 2012), mapped out how to move from a strategy and action plan to implementation. Additional developments have included a pan-Canada age-friendly network to connect age-friendly communities, the development of age-friendly training in private-sector companies and the creation of Age-Friendly Business Guides in specific locales (for example in 2016, the Age Friendly London Network in London, Ontario created the *Age Friendly Business Resource Guide*; Age Friendly London 2016). Although analysis of tourism and leisure within the WHO age-friendly locales demonstrates only very limited progress in transforming the visitor economy, there have been some promising initiatives. For example, in Seochu-gu, a residential district of Seoul, South Korea, an action plan to create age-friendly outdoor spaces was developed, encouraging intragenerational use as well as by people with dementia. To champion this, local authorities were charged with engaging civil society buddies and older-person associations to contribute to the planning and policymaking in order to address specific barriers (e.g. single households, isolation and predominantly home-based leisure). In Canada, Age-Friendly Ottawa (2017) produced the report *How Age-Friendly Is Ottawa?*, which found that 20% of the city's population will be over 60 by 2035, double the 2015 rate of 116,500 'seniors'. Key findings were that among ageing residents:

- 89% were satisfied with life.
- 79% found it easier to walk in the local neighbourhood for leisure than to go shopping.
- 70% had good mental health.
- 47% walked for 1–5 hours a week in the local community.

But problems associated with seniors were that:

- 31% walked for less than an hour a week.
- 21% had a fall in the last year, half of whom found it had made life more difficult.
- Some seniors faced elder abuse.

Age-Friendly Ottawa sought to enhance the walkability of neighbourhoods that would facilitate near-to-home leisure. For example, measures such as improving the safety and accessibility of leisure sites was implemented (also observed by Gibney *et al.* 2020). Worrying features were that 11% of ageing people had been involved in car crashes as pedestrians, and only 23% used public transport. More data monitoring was needed on initiatives to engage the ageing residents in civic volunteering and social participation. The project Age-Friendly Hamilton launched a three-part initiative in 2017; it included the projects Let's Get Moving, Let's Take the Bus and Let's Take a Walk. As part of this initiative, 18 recreational trails were designed; maps were then published and distributed to 400 older adults to promote active ageing. Age-Friendly Hamilton recruited 700 older residents as volunteers to help ensure outdoor spaces and public places were accessible. One of the few examples in which this has been applied to a business setting in Canada was in British Columbia, where

> an age-friendly business [focused] on making sure that older customers [were] safe, comfortable and respected by including such features as:
>
> * Non-slip surfaces, wider aisles and easily-opened doors;
> * Clear signage;
> * Places to sit and refresh; and
> * Patient, helpful staff
>
> (www2.gov.bc.ca)

In the UK, the Centre for Ageing Better supports the UK's network of 44 age-friendly communities. Clearly, this has particular salience for the visitor economy in relation to leisure because the WHO (2007b) guidance has led to a focus on outdoor environments, which many countries have adopted in a bid to boost physical and mental well-being. The emphasis has been on creating clean environments like green space, accessed by age-friendly pavements, safe pedestrian crossings and secure cycle paths, and supported by age-friendly buildings (which encompass visitor economy infrastructure such as visitor attractions, hotels and airports). But the evidence to date is that the creation and effectiveness of age-friendly environments needs to be assessed, and the extent to which age-friendly principles have been embedded in visitor economy business practices needs much greater exploration.

A highly contentious area of age-friendly infrastructure is toilet provision, given the link between continence issues and ageing. In the UK, the British Toilet Association (www.btaloos.co.uk) reiterates the importance of access to clean and hygienic public toilets. Yet between 2009 and 2010, 700 council-run toilets were closed due to budget cuts, although pay-to-use toilets and provision in cafes, hospitality establishments and shops have partially filled this gap. Frizzell (2019) estimates that between 1999 and 2019 the number of public conveniences was cut by 33%, while other providers have resorted to charging to offset their running costs. The exception to this was the UK's 20 busiest mainline stations managed by Network Rail, as charges for these were removed in 2019. For example, at London Liverpool Street, 1.3 million of the 47 million passengers who pass through annually use the facilities. To put this in perspective, there are around 10,000 public toilets in the UK on the Great British Public Toilet Map, which helps people to locate public conveniences when travelling. Such provision is a major necessity for any age-friendly environment if one expects ageing leisure activity to occur at any location beyond a short walk from home and back. As Page *et al.* (2015) found in the case of dementia, poor toilet provision can actually deter people from engaging in leisure travel, although this also has a wider significance for the whole ageing population in general, as this is a group that typically experiences greater frequency in terms of requiring the use of toilet facilities. Knight and Bichard (2011) argue that the issue is seen as distasteful, embarrassing or even funny and is often marginalised by policymakers, although as the British Toilet Association point out, access to a public toilet is a human right under the UN's Sustainable Development Goals. Indeed, this is not an issue wholly for older people, and good provision benefits the whole population.

Buffel *et al.* (2012) point to the weaknesses in the active-ageing paradigm as a way to enhance quality of life that is inherent in the WHO model of age-friendly communities. The problems in many megacities, where urban living conditions are squalid and basic needs are not even met (water, sanitation, secure housing), mean that promoting active ageing or quality-of-life issues are secondary to meeting the immediate basic needs of the population. Even so, the broader development of age-friendly environments to facilitate visitor economy activity evokes the debates raised by Lefebvre (1991) on the right to the city. Many ageing people are overlooked in urban regeneration schemes. Handler (2014) highlighted that future leisure development will need a much greater focus on community action so as to audit these environments from both an architectural and planning perspective to ensure they are fit for an age-friendly world.

The EU estimates that 1.2 million tourism enterprises will need to provide accessible tourism services to meet future demand, but training is very ad hoc via EU-funded projects (e.g. IN-TOUR – Inclusive Tourism, funded by ERASMUS). Transforming tourism supply will require a radical rethink about destination development and adaptation to new forms of demand. Johnson and Finn (2017) suggested that technology might be used to engage ageing populations through Universal Design Principles, in order to make places more accessible for those with impaired language, vision, motor control, learning/speech or cognition. Awareness of how older people navigate, consume and experience age-friendly cities and destinations/environments is needed in order to facilitate the evaluation of perceptions, attitudes and experiences (Dikken *et al.* 2020).

Implementation of age-friendly principles in the visitor economy

Sánchez-González *et al.* (2020) reviewed how age-friendly cities achieve active ageing, observing that environmental interventions help to reduce risk within the built and natural environment (e.g. slips, trips an falls) and psychosocial interventions to promote behaviour change and participation. Yet these interventions had limited success in the changing lifestyles of older people.

Alongside introducing Universal Design Principles around walking and walkability of communities, the concept of mobility as a service (MaaS) offers a flexible approach to mobility on demand as an alternative to the car in older age, as around one-third of an ageing population have unmet travel needs in cities (Luiu *et al.* 2018). This would build upon the health benefits of mobility to engage in leisure and tourism activity, through walking and seamless mobility (Musselwhite *et al.* 2015). Steels's (2015) review of age-friendly communities highlights that age-friendly interventions were process-driven and top-down initiatives with limited short-term outcomes. This means that the visitor economy remains largely hidden from focus in age-friendly communities. We suggest in Figure 7.3 that a distinct policy–practice conundrum exists where three stakeholders (the public sector who set the policy framework for age-friendly communities; the trade associations and organisations which manage the visitor economy; and the businesses involved in these activities) form a triangle; all three segments of that triangle need to be coordinated by means of a communication strategy to promote the opportunities of an age-friendly visitor environment.

Taking a lead from the existing research and initiatives, developing an age-friendly visitor economy would have to be based on the assumption that a public-sector-led, often top-down compliance approach is perhaps less able than a bottom-up approach to engage the visitor economy, with its preponderance of small and micro businesses. The assumption in existing age-friendly initiatives,

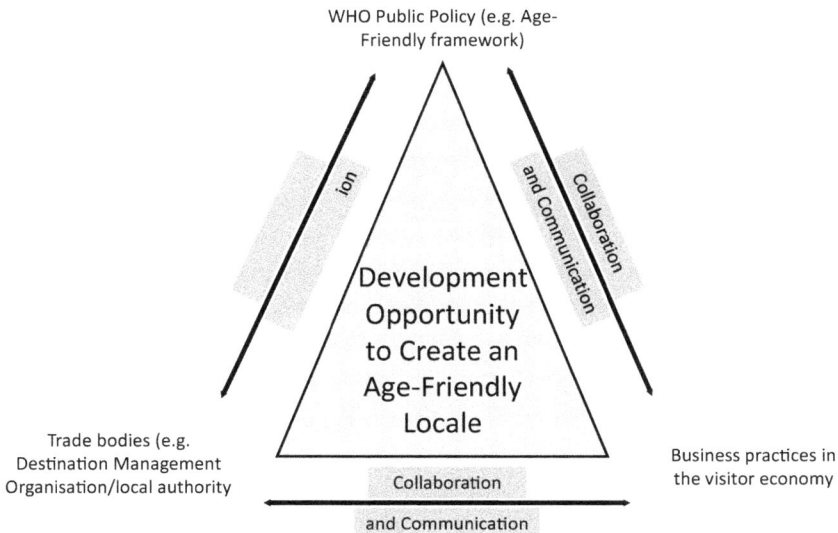

Figure 7.3 The opportunity to develop an age-friendly visitor environment: the policy–practice conundrum

from a policy perspective, is that people are altruistic, see the benefits that others champion and so compliance will achieve some degree of implementation. Many of the policy assumptions on achieving change are premised on the principle that a sea change can be made possible by introducing policies and selected interventions in two principal domains:

- *The individual*, to achieve behavioural changes (e.g. to secure engagement with an active or healthy ageing agenda) and
- *The environmental* (i.e. improvements to infrastructure and aspects of the built environment which individuals utilise, such as pavements, parks, buildings and facilities).

These two domains have seen public-sector investment since the initial United Nations Year of Older People in 1999 and the WHO Age-Friendly Cities Initiative in 2006. But these initiatives appear to have largely bypassed the visitor economy. For that sector, the missing piece of the jigsaw is businesses engagement with this agenda. Based on the experiences detailed in Chapter 6 associated with DF communities and environments, age-friendly projects have sought to make public spaces and buildings accessible and hazard-free as a key step forward towards age-friendliness, but broader engagement with individual businesses and visitor economy touchpoints remains an undeveloped area. The exception to this is the way airports have provided special assistance for travellers, with flat, navigable spaces and internationally agreed black signposting on a yellow background, which provides many cues for the visitor economy. The challenge is that many small- to medium-sized enterprises (SMEs) and individual businesses do not have the investment resources or ability to redesign their premises or environments, as Historic Royal Palaces (2017) outlined. They are also unlikely to engage with any public-sector-led initiative without a champion from a trade organisation or lead body like a destination management organisation (DMO). Building on our experience in working with dementia, we present the model in Figure 7.4 as one possible approach to illustrate how to make leisure and tourism spaces more age-friendly in a more holistic manner.

Based on the WHO model of transitioning towards a DF environment, Figure 7.4 is adapted from the model proposed by Connell *et al.* (2017) to illustrate the basic tenets of how an age-friendly visitor economy might address the inherent weakness within the WHO Age-Friendly framework. Figure 7.4 adopts a comprehensive approach rather than a piecemeal one, and is focused on specific elements of visitor economy touchpoints and four specific interconnected pillars:

- First, *place* comprises the resources, sites and venues that visitors use and that need to be assessed in terms of becoming age-friendly (we will return the theme of how to achieve this below). The provision of appealing visitor sites (i.e. attractions) is an essential drawcard for ageing visitors and specific adaptations are often needed (e.g. step-free access).

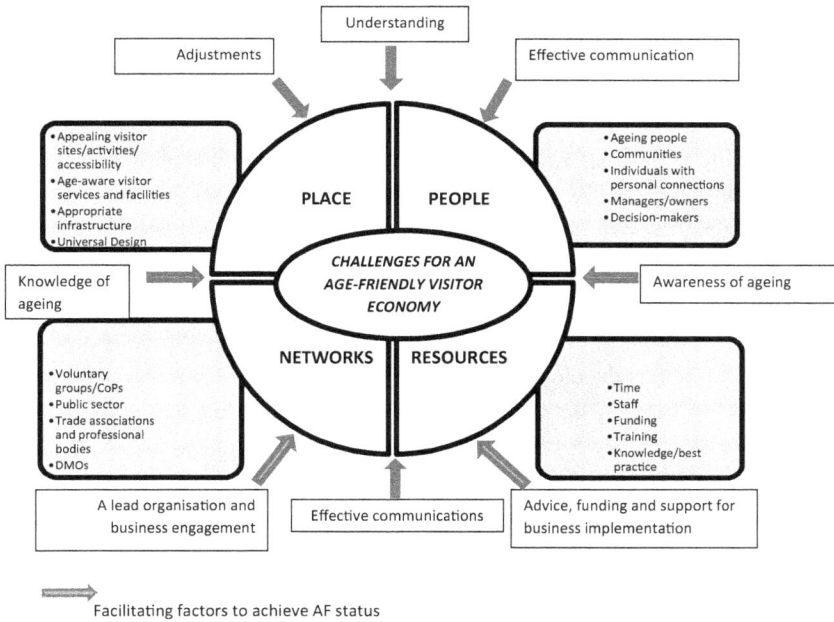

Figure 7.4 A framework for developing age-friendly visitor destination and leisure services

- Second, *networks* are an important component, as networking and collaboration can help integrate and access the expertise and knowledge required to understand visitors' design needs and challenges; these are often identified by walking the site or destination to illustrate barriers. At this point, examples of best practice and service blueprinting may help establish the type of age-friendly experience that a business can offer.
- Third, *people* are the basis of everything that happens in the visitor economy, from the ageing visitor, through the staff who interact with that visitor, to the people who manage the businesses. This also applies to the public sector, including those resources and places that are free to access such as parks and gardens. The people element is also important in championing age-friendly issues from developing the place, requiring networks to drive change. The cumulative effect of people in these networks is that they have a common purpose, like a CoP, and so are able to focus attention on a change agenda. People are also a 'resource' which dovetails with the fourth pillar to implement a more inclusive approach through building in best practice.
- Fourth, *resources* are another way of conceptualising and building the case for progressing age-friendly activity. For example, SMEs often claim they do not have the resources to engage with initiatives. Within tourism, perceived

barriers in the business community can probably only be broken down by the DMO recruiting a champion from the business community who is able to promote the innovations that are needed to make a difference, using low-cost interventions. This may also help businesses see how innovating might improve their bottom line.

As with the WHO's DF model, this process is one of gradual evolution for businesses as they pass through different stages. At the outset, our research evidence suggests that success is most likely where lead businesses champion the idea initially, coordinated by appropriate organisations such as DMOs to create an age-friendly itinerary to raise the profile of the offer. This would be enhanced through the increasing use of smart technology to connect visitors and leisure participants with accessible opportunities, tailored to their requirements.

Underpinning this development is the way in which the visitor economy makes its touchpoints age-friendly at two types of location:

• At a *destination* – or in terms of leisure, the *local leisure environment* – through the provision of age-friendly infrastructure and products/services which are connected (potentially through smart technology and social media) via itineraries and targeted communication strategies.
• *Within specific businesses*, showcasing the types of service they offer that are age-friendly in order to innovate and stimulate other businesses to follow suit.

As Figure 7.5 suggests, we advocate adopting the communities of practice model (CoP) as one low-cost but effective approach, in which the stimulus comes

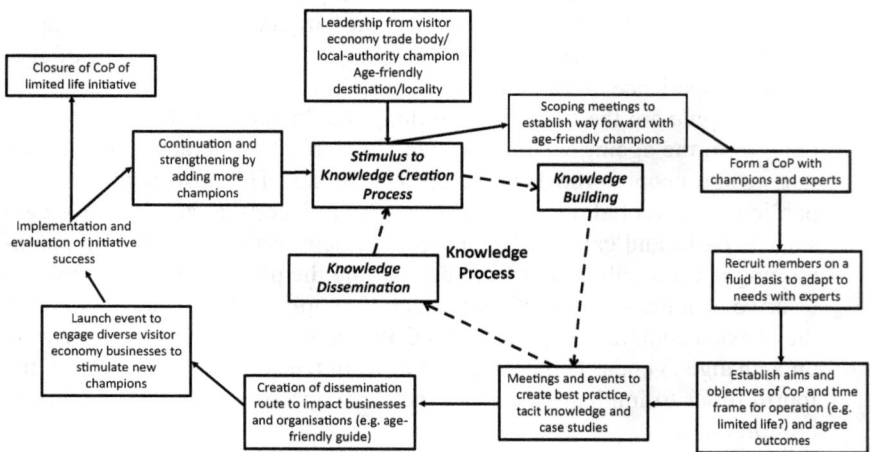

Figure 7.5 Creating age-friendly visitor environments using a community-of-practice approach

from a trade body, or other equally valid champion, which champions the idea and is able to recognise the needs of both destinations and businesses in creating a more age-friendly approach. One key action would be a preliminary blueprinting of the service offer at a destination and business level, supported by a summary case or guide for businesses in the same vein as the VisitEngland guide (2019) on dementia, which sought to encourage changes to business practices.

Blueprinting visitor economy service provision

Blueprinting services is a relatively recent development, emanating from services marketing, which is described as a multilayered process designed to help businesses adapt to customers' needs. The ideas in service blueprinting are influenced by operations research and its focus on business processes to achieve a greater focus on customer-centred design to help innovate and adapt. The origins of service blueprinting (SB) arose from the initial study by Shostack (1984), and the key principles of this process include:

*   Identifying the processes of services.
*   Isolating the 'fail points' in the business/its processes.
*   Setting the time frame to make an improvement.
*   Analysing profitability of the changes, which is critical in engaging businesses in becoming more age friendly.
*   The expected delivery outcomes.
*   The service process and service design issues and way these are managed.Developed from Shostack (1984)

The principles of service blueprinting have seen considerable development since Shostack's (1984) seminal work, as the notion of services marketing and management have developed a greater academic presence within business and management research. Key stimuli to this this development included Zeithaml and Bitner (1996), Lovelock *et al.* (2001), Zeithaml *et al.* (2006), Bitner *et al.* (2008) and Dirsehan (2020). The blueprinting process may also offer an opportunity for co-creation of age-friendly experiences in the visitor economy (Grönroos 2011). While developing the age-friendly destination may be a less desirable outcome than normalisation to make an entire society age friendly, it offers a return on investment for destinations able to offer the complete range of supporting experiences across the visitor economy. Such an approach would allow destinations to work towards a normalised agenda whereby age-friendly practice is championed and accommodated in a supportive and seamless manner rather than in an ad hoc style, or with businesses paying lip service to the issue.

And finally …

This book has demonstrated that before we make too many assumptions about ageing consumers and visitors as a minoritised group, we need to recognise that ageing is complex human process that affects people at different rates and in different ways. We have shown that the conceptualisation of ageing people as a homogenous group is not a meaningful assumption to make, and each chapter has shown the diverse characteristics associated with the ageing process, and how these interact with specific health conditions (it is worth noting that some health conditions have not been reviewed in this book, for example multiple sclerosis, which also has devastating impacts when combined with ageing). Factors such as individual life histories, the impact of retirement on leisure and the complex changes occurring around lengthening life expectancy, and the rise of multigenerational families make any analysis of ageing–visitor economy relationships complex. Trying to unpick the implications of these complex interrelated issues for the visitor economy is challenging, given the paucity of information, and the diversity of public-sector initiatives and interventions on ageing. While some specific areas of activity such as the interest in dementia and in making environments DF have surfaced in the last decade, these are only just emerging as visitor economy issues. Perhaps this book is ahead of its time, as there is currently a theory-implementation gap on ageing and the visitor economy that will not be closed until the researcher–policymaker–practitioner nexus is fostered in the way the DF agenda has been approached.

We have sought to raise and explore these themes in depth in this book in order to establish the groundwork for greater interdisciplinarity in the way ageing is framed in research agendas in relation to the visitor economy. We do not expect this to be anything short of a mountain to climb, especially in terms of implementation, but one positive sign is the growing interest in age-friendly communities. We need to be able to show a well-argued business case as to why the visitor economy should become more age friendly, using evidence and case studies of best practice in the journey towards normalisation. There is still a great deal of progress needed as attitudes toward older age range from ambivalence to ageism in many societies: we need to change the mindset and behaviours towards an ageing population. To end on a positive note, the visitor economy has an opportunity to reposition its offer so as to nurture an ageing population, building on the brand loyalty that some businesses have developed through their life course. To create new opportunities as part of an age-focused offer will require businesses to collaborate and cooperate to build more value from consumer interactions.

Appendix 1: Interview Questions

Part I. Ageing and tourism

1. On a personal level, what is your immediate response to the term 'ageing'?
2. What is the main focus of your company's business (If selling packages – what type of packages – escorted, self-organised, self-arranged)?
3. How are the tourism products/services you offer targeted at ageing travellers?
4. Approximately what proportion of your business is focused on ageing consumers?
5. How has the demand for products and services you provide changed in the last 5 years?
6. How do you communicate to consumers over the age of 55? Has this changed significantly with the rise of social media?
7. Do people aged over 55 feature prominently in your advertising of products? If so, how? What type of imagery do you tend to use (e.g. passive images, activities, groups, couples)?
8. What are the challenges for your organisation of working with more ageing consumers? Do they consume more of your time in terms of specific requirements and so add more cost to your operations? If so, can you list any specific examples?
9. Are there any specific adaptations, services or needs which you have to tailor for with this group? If so, can you outline some examples?
10. Does your organisation have any experience of ageing consumers travelling with people from other generations in their family (e.g. children and grandchildren)?
11. Do you place any restrictions on ageing travellers being required to disclose specific or multiple health conditions that could impact upon the experience of a holiday? If so, how is this managed?
12. What type of information do you provide to help people travel who have specific needs with (a) a disability, (b) a cognitive and mental issue such as dementia, (c) autism and other special needs?
13. In your product range, do you promote activities and exercise among the holiday experiences you offer?

14. Have you encountered any initiatives or organisations that seek to expand awareness of ageing? If so, can you outline the types of interactions you have had?
15. Can you please look into the future to 5–10 years away: How do you think the market for ageing tourism products will change? Will technology play a greater role in the way ageing consumers interact with your organisation?

Part II. Dementia and tourism and hospitality businesses

16. Do you think the tourism and hospitality sector sees dementia as an issue? Why do you think that?
17. If you responded 'yes' to the previous question, do you think that the tourism and hospitality sector is proactive in developing dementia-friendly initiatives compared with other sectors? If no, do you think it should be doing better? Why?

Part III. Your business/organisation

18. Would you say that your business/organisation is dementia aware (i.e. understands that some customers may be living with dementia) or dementia-friendly (i.e. has taken steps to address the needs of customers living with dementia)?
19. What do you think makes your business dementia aware/friendly? What steps have you taken in your business?
20. If you cannot say that organisation is dementia aware/friendly, is it something you are thinking about doing? Or something you do not really think applies to your business?
21. Do you feel there might be market opportunities for businesses that become dementia-friendly? Through enhanced customer service? Other aspects?
22. How do you think dementia-friendly initiatives might be better communicated to business?

Bibliography

Abeyratne, R. (1995) Proposal and guidelines for the carriage of elderly and disabled persons by air. *Journal of Travel Research*, 33(3):52–9.

Adam, I. (2019) Accommodators or non-accommodators? A typology of hotel frontline employees' attitude towards guests with disabilities. *International Journal of Hospitality Management*, 82:22–31.

Adam, I., Kumi-Kyereme, A. and Boakye, K. (2017) Leisure motivation of people with physical and visual disabilities in Ghana. *Leisure Studies*, 36(3):315–28.

Adams, J., Hayes, J. and Hopson, B. (1976) *Transition: Understanding and Managing Personal Change*. London: Martin Robson.

Adams, K., Leibbrandt, S. and Moon, H. (2010) A critical review of the literature on social and leisure activity and wellbeing in later life. *Ageing and Society*, 31(4):683–712.

Adams, K., Roberts, A. and Cole, M. (2011) Changes in activity and interest in the third and fourth age: associations with health, functioning and depressive symptoms. *Occupational Therapy International*, 18(1):4–17.

Adey, P., Bissell, D., Hannam, K., Merriman, P. and Sheller, M. (eds.) (2014) *The Routledge Handbook of Mobilities*. London: Routledge.

Agahi, N., Ahacic, K. and Parker, M. (2006) Continuity of leisure participation from middle age to old age. *The Journals of Gerontology: Series B*, 61(6): S340 – S346.

Age Cymru (2020) *Experiences of People Aged 50 or Over in Wales During the First Covid-19 Lockdown, and the Road to Recovery*. www.ageuk.org.uk/globalassets/age-cymru/documents/covid-19-survey/experiences-of-people-aged-50-or-over-in-wales-during-the-first-covid-19-lockdown-and-the-road-to-recovery – october-2020-eng.pdf.

Age Friendly London (2016) *Age Friendly Business Resource Guide*. www.information-london.ca/uploads/contentdocuments/afb%20resource%20guide_final.pdf

Age Friendly Ottawa (2017) *How Age-Friendly Is Ottawa? An Evaluation Framework to Measure the Age Friendliness of Ottawa*. Ottawa: The Council of Aging of Ottawa.

Age UK (2011) *Later Life in the UK*. London: Age UK.

Age UK (2014) *Housing in Later Life*. London: Age UK.

Age UK (2018) *All the Lonely People: Loneliness in Late Life*. London: Age UK.

Age UK (2019) *Poverty in Later Life*. London: Age UK.

Agovino, M., Casaccia, M., Garofalo, A. and Marchesano, K. (2017) Tourism and disability in Italy: limits and opportunities. *Tourism Management Perspectives*, 23:58–67.

Airault, R. and Valk, T. (2018) Travel-related psychosis (TrP): a landscape analysis. *Journal of Travel Medicine*, 25(1):tay054.

Aitchison, C. (2003) From leisure and disability to disability leisure: developing data, definitions and discourses. *Disability & Society*, 18(7):955–69.

Alén, E., Losada, N. and de Carlos, P. (2017) Profiling the segments of senior tourists throughout motivation and travel characteristics. *Current Issues in Tourism*, 20(14):1454–69.

Alén, E., Nicolau, J. L., Losada, N. and Domínguez, T. (2014) Determinant factors of senior tourists' length of stay. *Annals of Tourism Research*, 49:19–32.

Alley, D. and Chang, V. (2007) The changing relationship of obesity and disability 1988–2014. *Journal of the American Medical Association*, 298:2020–7.

Alley, D., Putney, N., Rice, M. and Bengtson, V. (2010) The increasing use of theory in social gerontology: 1990–2004. *Journals of Gerontology – Series B Psychological Sciences and Social Sciences*, 65B(5):583–90.

Altintas, E., Guerrien, A., Vivicorsi, B., Clément, E. and Vallerand, R. (2018) Leisure activities and motivational profiles in adaptation to nursing homes. *Canadian Journal on Aging/La Revue canadienne du vieillissement*, 37(3):333–44.

Alvesson, M. and Kärreman, D. (2001) Odd couple: making sense of the curious concept of knowledge management. *Journal of Management Studies*, 38(7):995–1018.

Alzheimer's Disease International (2016) *Dementia Friendly Communities: Global Developments*. London: Alzheimer's Disease International. www.alz.co.uk/adi/pdf/dfc-developments.pdf.

Alzheimer's Disease International and Global Coalition on Ageing (2017) *Dementia Innovation Readiness Index*. London: Alzheimer's Disease International and Global Coalition on Ageing. www.globalcoalitiononaging.com/data/uploads/documents/gcoa-adi-dementia-index.pdf.

Alzheimer's Society (2013) DFC survey. In Alzheimer's Society Report, *Building Dementia-Friendly Communities: A Priority for Everyone*. London: Alzheimer's Society.

Alzheimer's Society (2015) *How to help people with dementia: a guide for customer facing staff*. www.alzheimers.org.uk/sites/default/files/migrate/downloads/how_to_help_people_with_dementia_a_guide_for_customer-facing_staff.pdf

Alzheimer's Society (2016) Factsheet 474: Travelling and going on holiday. www.alzheimers.org.uk/sites/default/files/migrate/downloads/factsheet_travelling_and_going_on_holiday.pdf

Alzheimer's Society/British Standards Institute (2015) *Code of Practice for the Recognition of Dementia-Friendly Communities in England*. London: Alzheimer's Society/BSI.

Amadeus (2015) Future Traveller Tribes 2030. https://amadeus.com/documents/en/retail-travel-agencies/research-report/amadeus-future-traveller-tribes-2030-report.pdf

Amit, R. and Zott, C. (2012) Creating value through business model innovation. *MIT Sloan Management Review*, 53(3):41–9.

Ancient, C. and Good, A. (2014) *Considering people living with dementia when designing interfaces*. Lecture Notes in Computer Science Vol. 8520. Berlin: Springer Verlag, 113–123. https://doi.org/10.1007/978-3-319-07638-6_12

Andrews, G. (2020) Aging and health. In A. Kobayashi (ed.), *International Encyclopedia of Human Geography* (2nd edition). Oxford: Elsevier, 67–71.

Andrews, G., Cutchin, M., McCracken, K., Phillips, D. and Wiles, J. (2007) Geographical gerontology: The constitution of a discipline. *Social Science and Medicine*, 65(1):151–68.

Andriot, H. and Roumilhac V. (2010) Voyages et activités culturelles en institution pour personnes âgées [Travel and cultural activities in care homes for the elderly]. *Soins Gerontology*, 86 (Nov.-Dec.):35–8.

Arch, A. and Abou-Zhara, S. (2007) How web accessibility guidelines apply to design for the ageing population. In G. Eizmendi, J. Azkoitia and G. Craddock (eds.), *Challenges for Assistive Technology*. Amsterdam: IOS Press, 937–41.

Areheart, B. (2008) When disability isn't just right: the entrenchment of the medical model of disability and the Goldilocks dilemma. *Indiana Law Journal*, 83: 181.

Argyle, M. (1992) *The Social Psychology of Everyday Life*. London: Methuen.

Argyle, M. (2001) *The Psychology of Happiness*. London: Routledge.

Atchley, R. (1971) Retirement and leisure participation: continuity or crisis? *The Gerontologist*, 11(1):13–17.

Atchley, R. (1989) A continuity theory of normal aging. *The Gerontologist*, 29(2):183–90.

Atchley, R. (1999) *Continuity and Adaptation in Ageing: Creating Positive Experiences*. Baltimore, MD and London: John Hopkins University Press.

Atherton, M. (2017) *Deafness, Community and Culture in Britain: Leisure and Cohesion, 1945–95*. Manchester: Manchester University Press.

Austin, J. (2000) Strategic collaboration between nonprofits and businesses. *Nonprofit and Voluntary Sector Quarterly*, 29(1):69–97.

Austin, J. and Seitanidi, M. (2012) Collaborative value creation: A review of partnering between nonprofits and businesses: Part I. Value creation spectrum and collaboration stages. *Nonprofit and Voluntary Sector Quarterly*, 41(5):726–58.

AUT (2011) *The Tourism Needs of People with Hearing Loss*. Auckland: Auckland University of Technology. https://g3ict.org/index.php/actions/assetCount/download?id=4Oq5jaoXdmI0kts4lUTypzv1ME4fTuaBD0yOGZeQpKM%3D.

Avancini, A., Pala, V., Trestini, I., Tregnago, D., Mariani, L., Sieri, S., Krogh, V., Boresta, M., Milella, M., Pilotto, S. and Lanza, M. (2020) Exercise levels and preferences in cancer patients: A cross-sectional study. *International Journal of Environmental Research and Public Health*, 17(15):1–22.

Avramov, D., Maskova, M. and Europe, C. (2003) *Active Ageing in Europe*. Strasbourg: Council of Europe.

Bae, H., Jo, S.H. and Lee, E. (2021) Why do older consumers avoid innovative products and services? *Journal of Services Marketing*, 35(1):41–53.

Balderas-Cejudo, A., Patterson, I. and Leeson, G. (2019) Senior foodies: a developing niche market in gastronomic tourism. *International Journal of Gastronomy and Food Science*, 16:100152

Baloglu, S. and Shoemaker, S. (2001) Prediction of senior travelers' motorcoach use from demographic, psychological, and psychographic characteristics. *Journal of Travel Research*, 40(1), 12–18.

Banister, D. and Bowling, A. (2004) Quality of life for the elderly: the transport dimension. *Transport Policy*, 11(2):105–15.

Barclays (2015) *An Ageing Population: The Untapped Potential for Hospitality and Leisure Businesses*. London: Barclays.

Bardach, E. (2011) *A Practical Guide for Policy Analysis: The Eightfold Path to More Effective Problem Solving*. Thousand Oaks, CA: Sage.

Baron-Cohen, S., Lombardo, M., Auyeung, B., Ashwin, E., Chakrabarti, B. and Knickmeyer, R. (2011) Why are autism spectrum conditions more prevalent in males? *PLoS Biology*, 9(6):e1001081.

Baskaran, H. (2020) *Celebrating Active Ageing*. Chennai: Notion Press.

Battarra, R., Zuccaro, R. and Tremiterra, M. (2019) Smart mobility and elderly people. *TEME Journal of Land Use, Mobility and Environment*, 19:2342.

BBC (2016) Dementia now leading cause of death. www.bbc.co.uk/news/health-37972141

BBC News (2010) Cornwall NHS offers surfing as therapy for depression. 15 September. www.bbc.com/news/uk-england-cornwall-11314132

Beard, J. and Bloom, D. (2015) Towards a comprehensive public health response to population ageing. *The Lancet*, 385(9968):658–61.

Beaven, E. (2005) *Leisure, Citizenship and Working-Class Men in Britain, 1850–1945.* Manchester: Manchester University Press.

Bekö, G., Allen, J. G., Weschler, C. J., Vallarino, J. and Spengler, J. D. (2015) Impact of cabin ozone concentrations on passenger reported symptoms in commercial aircraft. *PLOS One*, 10(5):e0128454.

Bell, M. and Ward, G. (2000) Comparing temporary mobility with permanent migration. *Tourism Geographies*, 2(1):87–107.

Bell, S. L., Phoenix, C., Lovell, R. and Wheeler, B. W. (2015) Seeking everyday wellbeing: the coast as a therapeutic landscape. *Social Science and Medicine*, 142:56–67.

Beltrami, E. (1993) *Mathematical Models in the Social and Biological Sciences.* Boston, MA: Jones Bartlett.

Bengston, V., Burgess, E. and Parrott, T. (1997) Theory, explanation, and a third generation of theoretical development in sound gerontology. *Journal of Gerontology (Social Science)* 52:572–88.

Benioff, M., and Southwick, K. (2004) *Compassionate Capitalism: How Corporations can Make Doing Good an Integral Part of Doing Well.* Franklin, NJ: Careers Press.

Bentley, T. and Page, S. J. (2008) A decade of injury monitoring in the New Zealand adventure tourism sector: A summary risk analysis. *Tourism Management*, 29(5):857–69.

Berger, B. (1960) How long is a generation? *The British Journal of Sociology*, 11(1): 10–23.

Bering, J., Curtin, E. and Jong, J. (2017) Knowledge of deaths in hotel rooms diminishes perceived value and elicits guest aversion. *OMEGA – Journal of Death and Dying*, 79(3): 286–312.

Bernard, M., Bartlam, B., Sim, J. and Biggs, S. (2007) Housing and care for older people: life in an English purpose-built retirement village. *Ageing and Society*, 27(4): 555–78.

Bernard, M., Rickett, M., Amigoni, D., Munro, L., Murray, M. and Rezzano, J. (2015) Ages and stages: the place of theatre in the lives of older people. *Ageing & Society*, 35(6): 1119–45.

Bernini, C. and Cracolici, M. (2015) Demographic change, tourism expenditure and life cycle behaviour. *Tourism Management*, 47:191–205.

Berry, C. P. (2014) Austerity, ageing and the financialisation of pensions policy in the UK. *British Politics*, 11(1): 2–25.

Beutel, M., Brähler, E., Wiltink, J., Kerahrodi, J., Burghardt, J., Michal, M., Schulz, A., Wild, P., Münzel, T., Schmidtmann, I., Lackner, K., Pfeiffer, N., Borta, A. and Tibubos, A. (2019) New onset of depression in aging women and men: contributions of social, psychological, behavioral, and somatic predictors in the community. *Psychological Medicine*, 49(7):1148–55.

Bhatti, M. (2006) When I'm in the garden I can create my own paradise: homes and gardens in later life. *The Sociological Review*, 54(2):318–41.

Bickenbach, J. (2012) The International Classification of Functioning, disability and health and its relationship to disability studies. In N. Watson, A. Roulstone and C. Thomas (eds.), *Routledge Handbook of Disability Studies.* London: Routledge, 51–66.

Bieger, T. and Laesser, C. (2002) Swiss travel market – aspects of consumer behaviour in an aging travel market. *Tourism Review*, 57(4):23–7.

Biggs, S. (2008) Aging in a critical world: the search for generational intelligence. *Journal of Aging Studies*, 22(2):115–19.

Bimonte, V. and Faralla, V. (2012) Tourist types and happiness a comparative study in Maremma, Italy. *Annals of Tourism Research,* 39(4):1929–50.

Birchall, J. (2008) The 'mutualisation' of public services in Britain: a critical commentary. *Journal of Co-operative Studies*, 41(2):5–16.

Birks, J. (2016) *News and Civil Society: The Contested Space of Civil Society in UK media*. London: Routledge.

Birren, J., Cohen, G., Sloane, R., Lebowitz, B., Deutchman, D., Wykle, M. and Hooyman, N. (1992) *Handbook of Mental Health and Aging* (2nd edition). Oxford: Elsevier Science.

Bitner, M., Ostrom, A. and Morgan, F. (2008) Service blueprinting: a practical technique for service innovation. *California Management Review*, 50(3):66–94.

Blaikie, A. (1999) *Ageing and Popular Culture*. Cambridge: Cambridge University Press.

Blichfeldt, B. and Nicolaisen, J. (2011) Disabled travel: not easy, but doable. *Current Issues in Tourism*, 14(1), 79–102.

Bloice, L. and Burnett, S. (2016) Barriers to knowledge sharing in third sector social care: a case study. *Journal of Knowledge Management*, 20(1):125–45.

Blume, S. (2012) What can the study of science and technology tell us about disability? In N. Watson, A. Roulstone and C. Thomas (eds.), *Routledge Handbook of Disability Studies*. London: Routledge, 348–59.

Boksberger, P. and Laesser, C. (2009) Segmentation of the senior travel market by the means of travel motivations. *Journal of Vacation Marketing*, 15(4): 311–22.

Bolier, L., Haverman, M., Westerhof, G., Riper, H., Smit, F. and Bohlmeijer, E. (2013) Positive psychology interventions: a meta-analysis of randomized controlled studies. *BMC Public Health*, 13:119–38.

Booth, C. (1889) *Life and Labour of the People in London*. London: Macmillan.

Borges Tiago, M., Couto, J., Tiago, F. and Dias Faria, S. (2016) Baby boomers turning grey: European profiles. *Tourism Management*, 54:13–22.

Boulton-Lewis, G. and Tam, M. (eds.) (2012) *Active Ageing, Active Learning*. Dordrecht: Springer.

Bourne, R. R. A., Stevens, G. A., White, R. A., Smith, J. L., Flaxman, S. R., Price, H., Jonas, J. B., Keeffe, J., Leasher, J., Naidoo, K., Pesudovs, K., Resnikoff, S. and Taylor, H. R.Taylor, H. R. (2013) Causes of vision loss worldwide, 1990–2010: A systematic analysis. *The Lancet Global Health*, 1(6):e339 – e349.

Bowen, F., Newenham-Kahindi, A. and Herremans, I. (2010) When suits meet roots: the antecedents and consequences of community engagement strategy. *Journal of Business Ethics*, 95(2):297–318.

Bowling, A. and Gabriel, Z. (2007) Lay theories of quality of life in older age. *Ageing and Society*, 27(6):827–48.

Boxall, K., Nyanjom, J. and Slaven, J. (2018) Disability, hospitality and the new sharing economy. *International Journal of Contemporary Hospitality Management*, 30(1):539–56.

Boyer, G. and Schmidle, T. (2009) Poverty amongst the elderly in late Victorian Britain. *Economic History Review*, 62(2):249–78.

Bradshaw, S., Playford, E. and Riazi, A. (2012) Living well in care homes: a systematic review of qualitative studies. *Age and Ageing*, 41(4), 429–40.

Brandt, E. and Pope, A. (eds.) (1997) *Enabling America: Assessing the Rule of Rehabilitation Science and Engineering*. Washington, DC: National Academy Press.

Bray, R. and Raitz, V. (2001) *Flight to the Sun: The Story of the Holiday Revolution*. London: Continuum.

Bretschneider, S., Marc-Aurele, F. and Wu, J. (2005) 'Best Practices' research: a methodological guide for the perplexed. *Journal of Public Administration Research and Theory*, 15(2):307–23.

Briggs, A. (1963) *Victorian Cities*. London: Odhams Press.

British Library (n.d.) *Taking the Waters* (exhibition). https://web.archive.org/web/20200219210854/www.bl.uk/learning/langlit/texts/waters/takingthewaters.html

British Red Cross (2020) *Life after Lockdown: Tackling Loneliness among Those Left Behind*. London: Red Cross.

Browne, C. (1998) *Women, Feminism and Aging*. New York: Springer.

Brubaker, T. and Powers, E. (1976) The stereotype of 'old': a review and alternative approach. *Journal of Gerontology*, 31(4):441–7.

Buchanan, J., Husfeldt, J., Berg, T. and Houlihan, D. (2008) Publication trends in behavioral gerontology in the past 25 years: are the elderly still an understudied population in behavioral research? *Behavioral Interventions*, 23(1):65–74.

Buckley, R. (2020) Nature tourism and mental health: parks, happiness, and causation. *Journal of Sustainable Tourism*, 28(9):1409–24.

Buckley, R. and Brough, P. (2017) Economic value of parks via human mental health: an analytical framework. *Frontiers in Ecology and Evolution*, 5:16.

Buckley, R., Zhong, L. and Martin, S. (2021) Mental health key to tourism infrastructure in China's new megapark. *Tourism Management*, 82:104169

Buckner, S., Mattocks, C., Rimmer, M. and Lafortune, L. (2018) An evaluation tool for age-friendly and dementia friendly communities. *Working with Older People*, 22(1):48–58.

Buffel, T., Phillipson, C. and Scharf, T. (2012) Ageing in urban environments: developing age-friendly cities. *Critical Social Policy*, 32(4):597–617.

Buhalis, D. and Darcy, S. (eds.) (2011) *Accessible Tourism: Concepts and Issues*. Bristol: Channel View Publications.

Buhalis, D., Darcy, S. and Ambrose, I. (eds.) (2012) *Best Practice in Accessible Tourism: Inclusion, Disability, Ageing Population and Tourism*. Bristol: Channel View Publications.

Bullock, Charles C. and Mahon, Michael J. (2001) *Introduction to Recreation Services for People with Disabilities: A Person-Centred Approach*. Champaign, IL: Sagamore.

Buraway, M. (2005) Presidential Address: for Public Sociology. *American Sociology Review* 70:4–28.

Burgess, E. (ed.) (1959) *Aging in Western Societies: A Survey of Social Gerontology*. Chicago: University of Chicago Press.

Burki, T. K. (2019) UK health tourism for private cancer care. *The Lancet Oncology*, 20(3):334.

Burnett, J. and Baker, H. (2001) Assessing the travel-related behaviors of the mobility-disabled consumer. *Journal of Travel Research*, 40(1):4–11.

Burnett-Wolle, S. and Godbey, G. (2007) Refining research on older adults' leisure: implications of selection, optimization, and compensation and socioemotional selectivity theories. *Journal of Leisure Research*, 39(3):498–513.

Burns, C. (1932) *Leisure in the Modern World*. New York: Century and Co.

Burns, R. and Graefe, A. (2007) Constraints to outdoor recreation: exploring the effects of disabilities on perceptions and participation. *Journal of Leisure Research*, 39(1):156–181.

Butler, R. (1969) Age-ism: another form of bigotry. *Gerontologist*, 9(4):243–6.

Butler, R. (2006) Ageism. In R. Schulz (ed.), *The Encyclopedia of Ageing A – K*. New York: Springer, 41–2.

Buys, E. and Miller, E. (2007) The physical, leisure and social activities of very old Australian men living in a retirement village and the community. *Geriaction*, 25(2):15–19.

Cacioppo, J. and Patrick, W. (2008) *Loneliness: Human Nature and the Need for Social Connection*. New York: W. W. Norton.

Cajaiba-Santana, G. (2014) Social innovation: Moving the field forward. A conceptual framework. *Technological Forecasting and Social Change*, 82: 42–51.

Cameron, E and Green, M. (2004) *Making Sense of Change Management*. London: Kogan Page.

Canadian Hearing Society (1989) *Intercity Travel and the Deaf and Hard of Hearing Traveller: An Analysis of the Current State of Accessibility*, Vol. 9839. Montreal: Transport Canada.

Carmel, E. and Harlock, J. (2008) Instituting the 'third sector' as a governable terrain: partnership, procurement and performance in the UK. *Policy & Politics*, 36(2): 155–71.

Carneiro, M., Eusébio, C., Kastenholz, E. and Alvelos, H. (2013) Motivations to participate in social tourism programmes: a segmentation analysis of the senior market. *Anatolia*, 24(3): 352–66.

Carney, G. and Nash, P. (2020) *Critical Questions for Ageing*. Bristol: Policy Press.

Carstensen, L. (1992) Social and emotional patterns in adulthood: support for socioemotional selectivity theory. *Psychology and Aging*, 7:331–8.

Carstensen, L., Fung, H. and Charles, H. (2003) Socioemotional selectivity theory and the regulation of emotion in the second half of life. *Motivation and Emotion*, 27(2):103–23.

Cartwright, R. (1996) Travellers' diarrhoea. In S. Clift and S. J. Page (eds.), *Health and the International Tourist*. London: Routledge, 44–66.

Casey, B. (2012) The implications of the economic crisis for pensions and pension policy in Europe. *Global Social Policy*, 12(3):246–65.

Cassia, F., Castellani, P., Rossato, C. and Baccarani, C. (2020) Finding a way towards high-quality, accessible tourism: the role of digital ecosystems. *TQM Journal*, 33(1):205–21.

Cavan, R., Burgess, E., Havighurst, R. and Goldhamer, H. (1949) *Personal Adjustment in Old Age*. Chicago: Science Research Associates.

CDC (2020) *Yellow Book: Health Information for International Travel*. Oxford: Oxford University Press.

Cejudo, M. (2018) *Senior Tourism: Determinants, Motivations and Behaviour in a Globalized and Evolving Market*. Madrid: ESIC Editorial.

Centre for Ageing Better (2020) *Doddery But Dear? Examining Age-Related Stereotypes*. London: Centre for Ageing Better.

Centre for Economics and Business Research (2019) *The economic cost of dementia to English businesses – 2019 update: A report for Alzheimer's Society*. London: CEBR. www.alzheimers.org.uk/sites/default/files/2019-09/The%20economic%20cost%20 of%20dementia%20to%20English%20businesses%20-%20edited.pdf.

Chang, L. (2014) The relationship between nature-based tourism and autonomic nervous system function among older adults. *Journal of Travel Medicine*, 21(3):159–62.

Chase, K. (2009) *The Victorians and Old Age*. Oxford: Oxford University Press.

Chełmińska, K. and Jaremin, B. (2002) Travelling diabetics. *International Maritime Health*, 53(1–4):67–76.

Chen, B., Qiu, Z., Usio, N. and Nakamura, K. (2018) Tourism's impacts on rural livelihood in the sustainability of an aging community in Japan. *Sustainability*, 10(8):2896.

Chen, S. and Gassner, M. (2012) An investigation of the demographic, psychological, psychographic, and behavioral characteristics of Chinese senior leisure travelers. *Journal of China Tourism Research*, 8(2): 123–145.

Chen, S. and Shoemaker, S. (2014) Age and cohort effects: the American senior tourism market. *Annals of Tourism Research*, 48:58–75.

Chen, Y. and Li, X. (2018) Does a happy destination bring you happiness? Evidence from Swiss inbound tourism. *Tourism Management*, 65:256–66.

Chen, Y., Lehto, X. and Cai, L. (2013) Vacation and well-being: a study of Chinese tourists. *Annals of Tourism Research*, 42:284–310.

Chochinov, H. M. (2001) Depression in cancer patients. *Lancet Oncology*, 2(8):499–505.

Christou, P. and Simillidou, A. (2020) Tourist experience: the catalyst role of tourism in comforting melancholy, or not. *Journal of Hospitality and Tourism Management*, 42:210–21.

Chu, A. and Chu, R. (2013) Service willingness and senior tourists: knowledge about aging, attitudes toward the elderly, and work values. *Service Industries Journal*, 33(12):1148–64.

Chung, J. and Simpson, S. (2020) Social tourism for families with a terminally ill parent. *Annals of Tourism Research*, 84: 102813.

Cirella, G., Bąk, M., Kozlak, A., Pawłowska, B. and Borkowski, P. (2019) Transport innovations for elderly people. *Research in Transportation Business & Management*, 30:100381.

Clark, A., Fleche, S., Layard, R., Powdthavee, N. and Ward, G. (2018) *The Origins of Happiness: The Science of Wellbeing over the Life Course*. Princeton, NJ: Princeton University Press.

Clark, R., Burkhauser, R., Moon, M., Quinn, J. and Smeeding, T. (2004) *The Economics of an Aging Society*. Malden, MA: Blackwell Publishing.

Clarke, P., Ailshire, J., Bader, M., Morenoff, J. and House, J. (2008) Mobility disability and the urban built environment. *American Journal of Epidemiology*, 168(5), 506–13.

Cleaver, M., Muller, T., Ruys, H. and Wei, S. (1999) Tourism product development for the senior market, based on travel-motive research. *Tourism Recreation Research*, 24(1): 5–11.

Cloquet, I., Palomino, M., Shaw, G., Stephen, G. and Taylor, T. (2018) Disability, social inclusion and the marketing of tourist attractions. *Journal of Sustainable Tourism*, 26(2):221–37.

Coakes, E. and Smith, P. (2007) Developing communities of innovation by identifying innovation champions. *The Learning Organisation*, 14(1):74–85.

Cole, T. (1992) *The Journey of Life: A Cultural History of Aging in America*. Cambridge: Cambridge University Press.

Connell, J. and Page, S. J. (2019a) An exploratory study of creating dementia-friendly businesses in the visitor economy: Evidence from the UK. *Heliyon*, 5(4):e01471.

Connell, J. and Page, S. J. (2019b) Case study: Destination readiness for dementia-friendly visitor experiences: A scoping study. *Tourism Management*, 70:29–41.

Connell, J., Page, S. J., Sheriff, I. and Hibbert, J. (2017) Business engagement in a civil society: transitioning towards a dementia-friendly visitor economy. *Tourism Management*, 61:110–28.

Conrad, N., Judge, A., Tran, J., Mohseni, H., Hedgecott, D., Crespillo, A. *et al.* (2018) Temporal trends and patterns in heart failure incidence: A population-based study of 4 million individuals, *The Lancet*, 391: 10120.

Conti, D. and Burton, W. (1994) The economic impact of depression in a workplace. *Journal of Occupational Medicine*, 36:983–8.

Cooper, R. (2018). *What Is Civil Society? How Is the Term Used and What Is Seen to Be Its Role and Value (Internationally) in 2018?* Brighton: Institute of Development Studies.

Corbin, J. and Strauss, A. (2014) *Basics of Qualitative Research: Techniques and Procedures for Developing Grounded Theory*. London: Sage.

Corfield, A. and Paton, R. (2016) Investigating knowledge management: Can KM really change organisational culture? *Journal of Knowledge Management*, 20(1):88–103.

Cornwell, B., Laumann, E. and Schumm, L. (2008) The social connectedness of older adults: A national profile. *American Sociological Review*, 73:185–203.

Cossar, J., Reid, D., Fallon, R., Bell, E., Riding, M., Follett, E., Dow, B., Mitchell, S. and Grist, N. (1990) A cumulative review of studies of travellers, their experience of illness and the implications of these findings. *Journal of Infection*, 21: 27–42.

Coughlin, J. (2017) *The Longevity Economy*. New York: Public Affairs.

Countryside Recreation Research Advisory Group (1970) *Countryside Recreation Glossary*. London: Countryside Commission.

Cowdry, E. (1942) *Problems of Ageing* (2nd edition). Baltimore, MD: The Williams and Wilkins Company.

Cowdry, E. (ed.) (1939) *Problems of Ageing: Biological and Medical Aspects*. Baltimore, MD: The Williams and Wilkins Company.

Cox, G. and Thompson, N. (2021) *Death and Dying: Sociological Perspectives*. London: Routledge.

Crampton, A. (2009) *Global Aging: Emerging Challenges*. The Pardee Papers (6). www.bu.edu/pardee/files/2009/09/pardee_aging-6-global-aging.pdf

Crampton, J. and Eley, R. (2013) Dementia-friendly communities: what the project 'creating a dementia-friendly York' can tell us. *Working with Older People*, 17(2): 49–57.

Crampton, J., Dean, J. and Eley, R. (2012) *Creating a Dementia-Friendly York*. York: Joseph Rowntree Foundation.

Crawford, D. and Godbey, G. (1987) Reconceptualizing barriers to family leisure. *Leisure Sciences*, 9(2):119–27.

Crawford, D., Jackson, E. and Godbey, G. (1991) A hierarchical model of leisure constraints. *Leisure Sciences*, 13 (4):309–20.

Creswell, J. (2013) *Qualitative Inquiry and Research Design: Choosing among Five Approaches* (4th ed.). London: Sage.

Creswell, J. (2014) *A Concise Introduction to Mixed Methods Research*. London: Sage.

Crompton, J. (1979) Motivations for pleasure vacation. *Annals of Tourism Research*, 6(4):408–424.

Cumming, E. and Henry, W. (1961) *Growing Old*. New York: Basic.

Curl, A., Fitt, H. and Tomintz, M. (2020) Experiences of the built environment, falls and fear of falling outdoors among older adults: an exploratory study and future directions. *International Journal of Environmental Research and Public Health*, 17(4): 1224.

Currie, J. (1793) The effects of water, cold and warm as a remedy in fever and other diseases. *Annals of Medicine* (Edinburgh) 3:1–33.

Cutchin, M. (2009) Geographical gerontology: new contributions and spaces for development. *The Gerontologist* 49:440–4.

Dann, E., and Dann, W. (2012) Sightseeing for the sightless and soundless: tourism experiences of the deafblind. *Tourism, Culture and Communication*, 12(2):125–40.

Dann, G. (1981) Tourist motivation: an appraisal. *Annals of Tourism Research*, 8(2):187–219.

Darcy, S. (2003) 'Disability', in J. Jenkins and J Pigram (eds.), *Encyclopedia of Leisure and Outdoor Recreation*. London: Routledge, 114–18.

Darcy, S. (2010) Inherent complexity: Disability, accessible tourism and accommodation information preferences. *Tourism Management*, 31(6):816–26.

Darcy, S. (2012a) (Dis)embodied air travel experiences: Disability, discrimination and the affect of a discontinuous air travel chain. *Journal of Hospitality and Tourism Management*, 19(1):91–101.

Darcy, S. (2012b) Disability, access, and inclusion in the event industry: a call for inclusive event research. *Event Management*, 16(3):259–65.

Darcy, S., Cameron, B. and Pegg, S. (2010) Accessible tourism and sustainability: a discussion and case study. *Journal of Sustainable Tourism*, 18(4):515–37.

Darcy, S. and Dickson, T. (2009) A whole-of-life approach to tourism: the case for accessible tourism experiences. *Journal of Hospitality and Tourism Management*, 16(1):32–44.

Darcy, S., McKercher, B. and Schweinsberg, S. (2020) From tourism and disability to accessible tourism: a perspective article. *Tourism Review*, 75(1):140–4.

Darrat, M. and Flaherty, G. (2019) Retrospective analysis of older travellers attending a specialist travel health clinic. *Tropical Diseases, Travel Medicine and Vaccines*, 5(1):17.

Daruwalla, P. and Darcy, S. (2005) Personal and societal attitudes to disability. *Annals of Tourism Research*, 32(3):549–70.

Dattolo, A. and Luccio, F. (2016) A review of websites and mobile applications for people with autism spectrum disorders: towards shared guidelines. In *International Conference on Smart Objects and Technologies for Social Good*. Cham: Springer, 264–73.

Davidson, J. and Milligan, C. (2004) Embodying emotion sensing space: introducing emotional geographies. *Social & Cultural Geography*, 5(4), 523–32.

Davidson, J., Bondi, L. and Smith, M. (eds.),(2005) *Emotional Geographies*. Aldershot: Ashgate.

de Guzman, A., Magnayon, J., Manuel, P., Moratillo, C. and Lim, P. (2019) It takes two to tango: phenomenologizing Filipino Tour Guides experiences of aging foreign tourists' mindfulness and involvement. *Educational Gerontology*, 45(11):657–69.

Deery, M., Jago, L. and Mair, J. (2011) Volunteering for museums: the variation in motives across volunteer age groups. *Curator: The Museum Journal*, 54(3): 313–25.

Del Val, M. and Fuentes, (2003) Resistance to change: a literature review and empirical study. *Management Decision*, 41(2):148–55.

De Leo, D., Neulinger, K. and Cantor, C. (2001) *Ageing and Suicide*. http://hdl.handle.net/10072/7266

DeLeon, D. (2008) The historical roots of the field. In M. Moran, M. Rein and R. Goodin (eds.), *The Oxford Handbook of Public Policy*. Oxford: Oxford University Press, 39–57.

Deloitte and Oxford Economics (2008) *The Economic Case for the Visitor Economy: Final Report*. London: VisitBritain.

Dembowski, F. (2013) The roles of benchmarking, best practices and innovation in organisations' effectiveness. *The International Journal of Organizational Innovation*, 5(3):6–20.

den Hoed, W. (2020) Where everyday mobility meets tourism: an age-friendly perspective on cycling in the Netherlands and the UK. *Journal of Sustainable Tourism*, 28(2):185–203.

Dening, T. and Milne, A. (eds.) (2011) *Mental Health and Care Homes*. Oxford: Oxford University Press.

Dent, E. and Goldberg, S. (1999) Challenging 'resistance to change'. *The Journal of Applied Behavioral Science*, 35(1):25–41.

Department for Business, Energy and Industrial Strategy (2017) *Building our Industrial Strategy*. London: Department for Business, Energy and Industrial Strategy. https://assets.publishing.service.gov.uk/government/uploads/system/uploads/attachment_data/file/664563/industrial-strategy-white-paper-web-ready-version.pdf

Department of Health (2015) *Prime Minister's Challenge on Dementia 2020*. London: Department of Health.

Destination Canada (2018) *Unlocking the Potential of Canada's Visitor Economy*. www.destinationcanada.com/en/news/unlocking-the-potential-of-Canadas-visitor-economy

Deuschl, G., Beghi, E., Fazekas, F., Varga, T., Christoforidi, K. A., Sipido, E., Bassetti, C., Vos, T. and Feigin, V. (2020) The burden of neurological diseases in Europe: an analysis for the Global Burden of Disease Study 2017. *The Lancet Public Health*, 5(10):e551 – e567.

Devile, E. and Kastenholz, E. (2018) Accessible tourism experiences: the voice of people with visual disabilities. *Journal of Policy Research in Tourism, Leisure and Events*, 10(3):265–85.

Dickson, T., Misener, L. and Darcy, S. (2017). Enhancing destination competitiveness through disability sport event legacies: developing an interdisciplinary typology. International *Journal of Contemporary Hospitality Management*, 29(3):924–46.

Dickson, T. J., Darcy, S. and Benson, A. (2018) Volunteers with disabilities at the London 2012 Olympic and Paralympic Games: who, why, and will they do it again. *Event Management*, 21(3):301–18.

Dikken, J., van den Hoven, R., van Staalduinen, W., Hulsebosch-Janssen, L. and van Hoof, J. (2020) How older people experience the age-friendliness of their city: development of the age-friendly cities and communities questionnaire. *International Journal of Environmental Research and Public Health*, 17(18):1–24.

Dioko, L. and Harrill, R. (2019) Killed while traveling – Trends in tourism-related mortality, injuries, and leading causes of tourist deaths from published English news reports, 2000–2017. *Tourism Management*, 70:103–23.

Dirsehan, T. (ed.),(2020) *Managing Customer Experiences in an Omnichannel World: Melody of Online and Offline Environments in the Customer Journey*. Bingley: Emerald Publishing Limited.

Dixon, A. (2020) *The Age of Ageing Better: A Manifesto for our Future*. London: Green Tree.

Djernes, J. (2006) Prevalence and predictors of depression in populations of elderly: a Review. *Acta Psychiatrica Scandinavica*, 113(5):372–87.

Doheny, S. and Jones, I. (2021) What's so critical about it? An analysis of critique within different strands of critical gerontology. *Ageing and Society*, 41(10): 1–21.

Domínguez Vila, T., Alén González, E. and Darcy, S. (2018) Website accessibility in the tourism industry: an analysis of official national tourism organization websites around the world. *Disability and Rehabilitation*, 40(24):2895–2906.

Domínguez Vila, T., Alén González, E. and Darcy, S. (2019) Accessible tourism online resources: a Northern European perspective. *Scandinavian Journal of Hospitality and Tourism*, 19(2):140–156.

Domínguez Vila, T., Alén González, E. and Darcy, S. (2020) Accessibility of tourism websites: the level of countries' commitment. *Universal Access in the Information Society*, 19(2):331–46.

Dowd, J. (1975) Aging as exchange: a preface to theory. *Journal of Gerontology*, 30:584–94.

Dredge, D. (2016) Are DMOs on a path to redundancy? *Tourism Recreation Research*, 41(3): 348–53.

Dubé, L., Bechara, A. Dagher, A., Drewnowski, A., Lebel, J., James, P. and Yada, R. (eds.) (2010) *Obesity Prevention*. San Diego: Academic Press.

Dubois, Y., Ravalet, E., Vincent-Geslin, S. and Kaufmann, V. (2015) Motility and high mobility. In G. Viry and V. Kaufmann (eds.), *High Mobility in Europe: Work and Personal Life*. Basingstoke: Palgrave Macmillan, 101–28.

Ducarme, J. (2020) COVID-19 is making America's loneliness epidemic even worse. *Time Magazine*, May 8, https://time.com/5833681/loneliness-covid-19/.

Duffield, B. (1984) The study of tourism in Britain – a geographical perspective. *GeoJournal*, 9(1):27–35.

Duggan, S., Blackman, T., Martyr, A. and Van Schaik, P. (2008) The impact of early dementia on outdoor life: a 'shrinking world'? *Dementia*, 7(2):191–204.

Dumazedier, J. (1967) *Toward a Society of Leisure*. Translated by Stewart E. McClure. New York: Free Press.

Dupuis, S. and Smale, B. (1995) An examination of the relationship between psychological well-being and depression and leisure activity among older adults. *Loisir & Societe/ Society and Leisure*, 18:67–92.

Durie, A. (2006) *Water is Best: The Hydros and Health Tourism in Scotland 1840–1940*. Edinburgh: John Donaldson Publisher Ltd.

Durko, A. and Petrick, J. (2013) Family and relationship benefits of travel experiences: a literature review. *Journal of Travel Research*, 52(6):720–30.

Dykstra, P. (2009) Older adult loneliness: myths and realities. *European Journal of Ageing*, 6(2):91–100.

Eades, M., Lord, K. and Cooper, C. (2016) 'Festival in a Box': development and qualitative evaluation of an outreach programme to engage socially isolated people with dementia. *Dementia*, 17(7):896–908.

Eberts, S. (2017) Tips for traveling with hearing loss. *The Hearing Journal*, 70(9):42.

Economic and Social Research Council (ESRC) (2021) Healthy ageing challenge. Swindon: UKRI. www.ukri.org/our-work/our-main-funds/industrial-strategy-challenge-fund/ageing-society/healthy-ageing-challenge /.

Edwards, M. (ed.) (2013) *The Oxford Handbook of Civil Society*. Oxford: Oxford University Press.

Ehrenberg, J. (2013) The history of civil society ideas. In M. Edwards (ed.), *The Oxford Handbook of Civil Society*. Oxford: Oxford University Press, 15–28.

Eichhorn, V., Miller, G., Michopoulou, E. and Buhalis, D. (2008) Enabling access to tourism through information schemes? *Annals of Tourism Research*, 35(1):189–210.

Ekerdt, D. and Koss, C. (2016) The task of time in retirement. *Ageing and Society*, 36(6):1295–1311.

Elvidge, N. (1973) The dynamics of population growth in the Central Coast of New South Wales. *Australian Geographer*, 12:226–36.

Engeland, J., Kittelsaa, A. and Langballe, E. (2018) How do people with intellectual disabilities in Norway experience the transition to retirement and life as retirees? *Scandinavian Journal of Disability Research*, 20(1):72–81.

Erevelles, N. (2011) *Disability and Difference in Global Contexts: Enabling a Transformative Body Politic*. London: Palgrave Macmillan.

Erikson, E. (1963) *Childhood and Society*. New York: Norton.

Estes, C., Swan, J. and Gerard, L. (1982) Dominant and competing paradigms in gerontology: towards a political economy of ageing. *Ageing and Society*, 2(2):151–64.

Etzioni, A. (2000) *The Third Way to a Good Society*. London: Demos.

Eusébio, C., Carneiro, M., Kastenholz, E. and Alvelos, H. (2017) Social tourism programmes for the senior market: a benefit segmentation analysis. *Journal of Tourism and Cultural Change*, 15(1):59–79.

Evans, M., Shickle D. and Morgan M. (2001) Travel illness in British package holiday tourists: prospective cohort study. *Journal of Infection*, 43(2):140–7.

Ezza, A., Ludovico, M. and Giovanelli, L. (2019) Assessing the impact of tourism on hospitals' performance in a coastal destination. *Tourismos: An International Multidisciplinary Journal of Tourism*, 1 (14):55–76.

Faranda, W. and Schmidt, S. (2000) Segmentation and the senior traveller: implications for today's and tomorrow's aging consumer. *Journal of Travel and Tourism Marketing*, 8(2):3–27.

Farmaki, A. and Stergiou, D. (2019) Escaping loneliness through Airbnb host – guest interactions. *Tourism Management*, 74:331–3.

Fastame, M., Hitchcott, P. and Penna, M. (2018) The impact of leisure on mental health of Sardinian elderly from the 'blue zone': evidence for ageing well. *Aging Clinical Experience Research*, 30:169–80.

Faulkner, A., Harding, T., Miller, C., Davies, P. and McNair, C. (2020) Tourism and the Highlands: a cross-sectional study on trauma and orthopaedic service use by tourists in 2017. *The Surgeon*, 19 (3):162–6.

FCO (Foreign and Commonwealth Office) (2013) *Travel and Mental Health.* London: Foreign and Commonwealth Office. www.gov.uk/guidance/foreign-travel-advice-for-people-with-mental-health-issues.

Feiler, T., Hordern, J. and Papanikitas, A. (eds.) (2018) *Marketisation, Ethics and Healthcare: Policy, Practice and Moral Formation.* London: Routledge.

Felkai, P. and Kurimay, T. (2017) Patients with mental problems – the most defenseless travellers. *Journal of Travel Medicine*, 24(5):tax005.

Felkai, P., Marcolongo, T. and Van Aswegen, M. (2020) Stranded abroad: a travel medicine approach to psychiatric repatriation. *Journal of Travel Medicine*, 27(2):taaa013.

Ferguson, A. (1787) *An Essay on the History of Civil Society* (5th edition). London: Cadell. https://oll.libertyfund.org/titles/1428

Ferrari, A., Charlson, F., Norman, R., Patten, S., Freedman, G., Murray, C., Vos, T. and Whiteford, H. (2010) Burden of depressive disorders by country, sex, age, and year: findings from the global burden of disease study, *PLOS Medicine*, 10(11):e1001547.

Ferrer, J., Sanz, M., Ferrandis, E., McCabe, S. and García, J. (2016) Social tourism and healthy ageing. *International Journal of Tourism Research*, 18(4):297–307.

Field, M. and Jette, A. (eds.) (2009) *The Future of Disability in America.* Washington, DC: National Academic Press.

Figueiredo, E., Eusebio, C. and Kastenholz, E. (2012) How diverse are tourists with disabilities? A pilot study on accessible leisure tourism experiences in Portugal. *International Journal of Tourism Research*, 14(6):531–50.

Filep, S. and Bereded-Samuel, E. (2012) Holidays against depression? An Ethiopian Australian initiative. *Current Issues in Tourism*, 15(3):281–5.

Filep, S. and Laing, J. (2018) Trends and directions in tourism and positive psychology. *Journal of Travel Research*, 58(3):343–54.

Filimonau, V. and Brown, L. (2018) 'Last hospitality' as an overlooked dimension in contemporary hospitality theory and practice. *International Journal of Hospitality Management*, 74: 67–74.

Finkel, R. and Dashper, K. (2020) Accessibility, diversity and inclusion in events. In S. J. Page and J. Connell (eds.), *The Routledge Handbook of Events* (2nd edition). London: Routledge, 475–90.

Finkel, R., Sharp, B. and Sweeney, M. (2019) Introduction. In R. Finkel, B. Sharp and M. Sweeney (Eds.), *Accessibility, Inclusion and Diversity in Critical Event Studies.* London: Routledge, 1–16.

Firth, L. and Mellor, D. (2007) Dilettantism in investigating the impact of the internet on the wellbeing of the elderly. *Quality & Quantity*, 43:185–96.

Fitzpatrick, T. (2018) Play, leisure activities, cognitive health, and quality of life among older cancer survivors. In T. Fitzpatrick (ed.), *Quality of Life Among Cancer Survivors: Challenges and Strategies for Oncology Professionals and Researchers.* New York: Springer International Publishing, 7–22.

Flaherty, G., Rossanese, A., Steffen, R. and Torresi, J. (2018) A golden age of travel: advancing the interests of older travellers. *Journal of Travel Medicine*, 25(1): tay088.

Fleetwood, C. (2020) Leading demographer says purpose is defining modern-day travel. *Travel Weekly*, 23 January. www.travelweekly.com.au/article/leading-demographer-says-purpose-is-defining-modern-day-travel/

Foley, N. and Rhodes, C. (2019) *Tourism: Statistics and Policy. Briefing Paper Number 06022*, 24 September 2018. Westminster: House of Commons Library.

Formosa, M. (ed.) (2019) *The University of the Third Age and Active Ageing: European and Asian-Pacific Perspectives*. Cham: Springer Nature.

Foster, L. (2018) Active ageing, pensions and retirement in the UK. *Journal of Population Ageing*, 11(2):117–32.

Fougeyrollas, P. (1989) Les implications de la diffusion de la Classification internationale des handicaps sur les politiques concernant les personnes handicappés. *World Health Statistical Quarterly*, 42:281–8.

Fougeyrollas, P. (1995) Documenting environmental factors for preventing the handicap creation process: Quebec contributions relating to ICIDH and social participation of people with functional differences. *Disability and Rehabilitation*, 17(3–4):145–60.

Fougeyrollas, P., Cloutier, R., Bergeron, H., Cote, J. and St Michel, G. (1998) *The Quebec Classification: Disability Creation Process*. Quebec: International Network on Disability Creation Process (INDCP/CSICIDH).

Franchetti, J. and Page, S. J. (2008) Entrepreneurship and innovation in tourism: public sector experiences of innovation activity in tourism in Scandinavia and Scotland. In J. Ateljevic and S. J. Page (eds.), *Tourism and Entrepreneurship*, Oxford: Butterworth-Heinemann, 107–30.

Fraser, J., Clayton, S., Sickler, J. and Taylor, A. (2009) Belonging at the zoo: retired volunteers, conservation activism and collective identity. *Ageing & Society*, 29(3): 351–68.

Freund, A. (2020) The bucket list effect: why leisure goals are often deferred until retirement. *American Psychologist*, 75(4): 499–510.

Freund, D., Cerdan Chiscano, M., Hernandez-Maskivker, G., Guix, M., Iñesta, A. and Castelló, M. (2019) Enhancing the hospitality customer experience of families with children on the autism spectrum disorder. *International Journal of Tourism Research*, 21(5):606–14.

Friends of the Elderly (2014) *The Future of Loneliness*: *Facing the Challenge of Loneliness*. London: Future Foundation.

Fries, J. (1980) Aging, natural death, and the compression of morbidity. *New England Journal of Medicine*, 303:130–5.

Frizzell, N. (2019) Wee demand action! What can we do about Britain's public toilet shortage? *The Guardian*, 15 January. www.theguardian.com/lifeandstyle/shortcuts/2019/jan/15/wee-demand-action-what-can-we-do-about-britains-public-toilet-shortage

Froggatt, K., Davies, S. and Meyer, J. (eds.) (2009) *Understanding Care Homes*. London: Jessica Kingsley Publishers.

Frost, I., Van Boeckel, T., Pires, J., Craig, J. and Laxminarayan, R. (2019) Global geographic trends in antimicrobial resistance: the role of international travel. *Journal of Travel Medicine*, 26(8):taz036.

Fullagar, S. (2008) Leisure practices as counter-depressants: emotion-work and emotion-play with women's recovery from depression. *Leisure Sciences*, 30(1):35–52.

Fullagar, S. and Owler, K. (1998) Narratives of leisure: recreating the self. *Disability and Society*, 13(3):441–50.

Funck, C. (2008) Ageing tourists, ageing destinations: Tourism and demographic change in Japan. In F. Coulmas, H. Conrad, A. Schad-Seifurt and G. Vogt (eds.), *The Demographic Challenge: A Handbook about Japan*. Leiden: Brill, 579–98.

Gabbay, S. and Wahler, J. (2002) Lesbian aging: review of a growing literature. *Journal of Gay and Lesbian Social Services*, 14(3):1–21.

Gallistl, V. and Nimrod, G. (2020) Media-based leisure and wellbeing: a study of older internet users. *Leisure Studies*, 39(2):251–65.

Gardner, J. (ed.) (1999) *Fundamentals of Feminist Gerontology*. New York: Haworth Press.

Gatrell, P., Elliott, P. and Williams, A. (2012) *Mobilities and Health*. Aldershot: Ashgate Publishing Limited.

Gaugler, J. (2014) Driving and other important activities in older adulthood. *Journal of Applied Gerontology*, 35(6):579–82.

Gauthier, S., Mausbach, J., Reisch, T. and Bartsch. C. (2015) Suicide tourism: a pilot study on the Swiss phenomenon. *Journal of Medical Ethics*, 41(8):611–17.

Gautret, P., Gaudart, J., Leder, K., Schwartz, E., Castelli, F., Lim, P., Murphy, H., Keystone, J., Cramer, J., Shaw, M., Boddaert, J., von Sonnenburg, F. and Parola, P. (2012) Travel associated illness in older adults. *Journal of Travel Medicine*, 19(3):169–77.

Generations United (2011) *Family Matters: Multigenerational Families in a Volatile Economy*. Washington, DC: Generations United. www.gu.org

Genoe, M. and Dupuis, S. (2012) The role of leisure within the dementia context. *Dementia*, 13(1):33–58.

Genoe, M. and Singleton, J. (2006) Older men's leisure experiences across their lifespan. *Topics in Geriatric Rehabilitation*, 22(4):348–56.

Gerometta, J., Haussermann, H. and Longo, G. (2005) Social innovation and civil society in urban governance: strategies for an inclusive city. *Urban Studies*, 42(11):2007–21.

Gershon, H. and Gershon, D. (2000) Paradigms in aging research: a critical review and assessment. *Mechanisms of Ageing and Development*, 117(1–3):21–8.

Gesler, W. (1992) Therapeutic landscapes: medical issues in the light of the new cultural geography. *Social Science and Medicine*, 34: 735–46.

Gesler, W. (1993) Therapeutic landscapes: theory and a case study of Epidaurus, Greece. *Environment and Planning D*, 11:171–89.

Gething, L. (2000) Ageing with long-standing hearing impairment and deafness. *International Journal of Rehabilitation Research*, 23(3):209–15.

Gholipour, H., Tajaddini, R. and Nguyen, J. (2016) Happiness and inbound tourism. *Annals of Tourism Research*, 57:251–3.

Gibney, S., Zhang, M. and Brennan, C. (2020) Age-friendly environments and psychosocial wellbeing: a study of older urban residents in Ireland. *Aging and Mental Health*, 24(12):2022–33.

Gibson, H., Gibson, J. and Singleton, J. (eds.) (2012) *Leisure and Aging: Theory and Practice*. Champaign, IL: Human Kinetics.

Giddens, A. (1998) *The Third Way: The Renewal of Social Democracy*. Cambridge: Polity.

Gilbert, E. W. (1965) The holiday industry and seaside towns in England and Wales. *Festschrift Leopold G. Scheidl zum 60. Geburtstag*. Vienna, 235–47.

Gillovic, B., McIntosh, A., Darcy, S. and Cockburn-Wootten, C. (2018) Enabling the language of accessible tourism. *Journal of Sustainable Tourism*, 26(4):615–30.

Giuffrida, A. (2020) Venice gondola tours reduce capacity due to 'overweight tourists'. *The Guardian*, 21 July. www.theguardian.com/world/2020/jul/21/venice-gondola-tours-reduce-capacity-due-to-overweight-tourists

Glasgow, R. (2013) What does it mean to be pragmatic? Pragmatic methods, measures, and models to facilitate research translation. *Health Education and Behaviour*, 40(3):257–65.

Glover, P. and Prideaux, B. (2009) Implications of population ageing for the development of tourism products and destinations. *Journal of Vacation Marketing*, 15(1):25–37.

Glyptis, S. (1981) Leisure life-styles. *Regional Studies*, 15:311–26.

Godbey, G., Crawford, D. and Shen, X. (2010) Assessing hierarchical leisure constraints theory after two decades. *Journal of Leisure Research*, 42(1):111–34.

Goddin, G. (2002) Whose heritage railway is it? A study of volunteer motivation. *Japan Railway & Transport Review*, 32:46–9.

Goelitz, D., TrenKamp, L. and Panlus, P. (2017) Leisure activities in care homes: how do they relate to the well-being of elderly. In Z. Benkö, I. Modi and K. Tarko (eds.), *Leisure Studies in a Global Era*. Singapore: Palgrave Macmillan, 73–8.

Golant, S. (1972) *The Residential Location and Spatial Behaviour of the Elderly*. Chicago: Chicago University Press.

Golant, S. M. (1984) The geographic literature on ageing and old age: An introduction. *Urban Geography* 5:262–72.

González, A., Rodríguez, C., Miranda, M. and Cervantes, M. (2009) Cognitive age as a criterion explaining senior tourists' motivations. *International Journal of Culture, Tourism and Hospitality Research*, 3(2):148–64.

González, E., Sánchez, N. and Vila, T. (2017) Activity of older tourists: understanding their participation in social tourism programs. *Journal of Vacation Marketing*, 23(4): 295–306.

Gössling, S., Cohen, S. and Hibbert, J. (2018) Tourism as connectedness. *Current Issues in Tourism*, 21(14):1586–1600.

Graham, A., Budd, L., Ison, S. and Timmis, A. (2019) Airports and ageing passengers: a study of the UK. *Research in Transportation Business & Management*, 30:100380.

Grandahl, K., Ibler, K., Laier, G. and Mortensen, O. (2018) Skin cancer risk perception and sun protection behavior at work, at leisure, and on sun holidays: a survey for Danish outdoor and indoor workers. *Environmental Health and Preventive Medicine*, 23(1):47.

Gray, A. and Dowds, L. (2010) *All Our Futures: Attitudes to Age and Ageing in Ireland. Research Update*. Belfast: ARK.ac.uk.

Gray, D. E. (1994) Coping with autism: stresses and strategies. *Sociology of Health & Illness*, 16(3):275–300.

Green, L., Ottoson, J., García, C. and Hiatt, R. (2009) Diffusion theory and knowledge dissemination, utilization, and integration in public health. *Annual Review of Public Health*, 30(1):151–74.

Gregoric, M., Skryl, T. and Drk, K. (2019) Accessibility of tourist offer in Republic of Croatia to people with disabilities. *Journal of Environmental Management and Tourism*, 10(4):903–15.

Grönroos, C. (2011) Value co-creation in service logic: a critical analysis. *Marketing Theory*, 11(3):279–301.

Gruenberg, E. (1977) The failure of success. *Millbank Memorial Fund Quarterly* 55 (1):773.

Grundey, D. and Vilutyte, G. (2012) Development of the tourism sector in Lithuania: a focus on the 50+ sector. *Journal of International Studies*, 5(1):30–7.

Guinn, R. (1980) Elderly recreational vehicle tourists: motivations for leisure. *Journal of Travel Research*, 19(1):9–12.

Gullette, M. (2017) *Ending Ageism, or How Not to Shoot Old People*. New York: Rutgers University Press.

Gutiérrez, J. and García-Palomares, J. (2020) Transport and accessibility. In A. Kobayashi (ed.), *International Encyclopedia of Human Geography* (2nd edition). Oxford: Elsevier, 407–14.

Hägerstrand, T. (1970) What about people in regional science? *Papers in Regional Science*, 24(1):7–24.

Hall, C. M. (2015) On the mobility of tourism mobilities. *Current Issues in Tourism*, 18(1):7–10.

Hall, C. M. and Page, S. J. (1999) *The Geography of Tourism and Recreation: Environment, Place and Space*. London: Routledge.

Hall, E. (2020) Accessibility. In A. Kobayashi (ed.), *International Encyclopedia of Human Geography* (2nd edition). Oxford: Elsevier, 1–8.

Hall, G. (1922) *Senescence, the Last Half of Life*. London: G. Appleton and Company.

Hall, J. (1995) (ed.) *Civil Society: Theory, History, Comparison*. Cambridge: Polity Press.

Hall, S., Opio, D., Dodd, R. and Higginson, I. (2011) Assessing quality-of-life in older people in care homes. *Age and Ageing*, 40(4):507–12.

Hamed, H. (2013) Tourism and autism: an initiative study for how travel companies can plan tourism trips for autistic people. *American Journal of Tourism Management*, 2(1):1–14.

Handler, S. (2014) *An Alternative Age-Friendly Handbook*. Manchester: University of Manchester Library.

Handy, C. (1976) *Understanding Organisations*. Harmondsworth: Penguin.

Hansen, M. and Birkinshaw, J. (2007) The innovation value chain. *Harvard Business Review*, June, 1–13.

Hantke, N., Etkin, A. and O'Hara, R. (2020) *Handbook of Mental Health and Aging*. Oxford: Elsevier Science.

Harahousou Y. (2006) Leisure and Ageing. In C. Rojek, S. M. Shaw, A. J. Veal (eds.), *A Handbook of Leisure Studies*. London: Palgrave Macmillan.

Harper, C., Hendrickson, C., Mangones, S. and Samaras, C. (2016) Estimating potential increases in travel with autonomous vehicles for the non-driving, elderly and people with travel-restrictive medical conditions. *Transportation Research Part C: Emerging Technologies*, 72:1–9.

Harper, S. (2006) *Ageing Societies*. London: Hodder Education.

Harrison, E., McKeown, M. and O'Shea, T. (1971) Old age in Northern Ireland – A study of the elderly in a seaside town. *Economic and Social Review* 3:53–72.

Hartung, T. J., Brähler, E., Faller, H., Härter, M., Hinz, A., Johansen, C., Keller, M., Koch, U., Schulz, H., Weis, J. and Mehnert, A. (2017) The risk of being depressed is significantly higher in cancer patients than in the general population: prevalence and severity of depressive symptoms across major cancer types. *European Journal of Cancer*, 72:46–53.

Hartwell, H., Fyall, A., Willis, C., Page, S. J., Ladkin, A. and Hemingway, A. (2018) Progress in tourism and destination wellbeing research. *Current Issues in Tourism*, 21(16):1830–92.

Harvey, D. (1990) *The Condition of Postmodernity*. Oxford: Blackwell.

Harvey, D. (2007) *A Brief History of Neoliberalism*. Oxford: Oxford University Press.

Harvey, D. (2008) The Right to the City. *New Left Review* 53, September – October.

Hauk, N., Hüffmeier, J. and Krumm, S. (2018) Ready to be a silver surfer? A meta-analysis on the relationship between chronological age and technology acceptance. *Computers in Human Behavior*, 84:304–19.

Havighurst, R. (1961) Successful ageing. *The Gerontologist*, 1:8–13.

Havighurst, R., Neugarten, B. and Tobin, B. (1963) Disengagement, personality and life satisfaction in the later years. In P. Hansen (ed.), *Age with a Future*. Copenhagen: Munksgaard, 419–25.

Hay, B. (2015) Dark hospitality: Hotels as places for the end of life. *Hospitality and Society*, 5(2/3):233–48.

Haynes, A., Nathan, A., Dixon, H., Wakefield, M. and Dobbinson, S. (2020) Sun-protective clothing and shade use in public outdoor leisure settings from 1992 to 2019: results from cross-sectional observational surveys in Melbourne, Australia. *Preventive Medicine*, 139, 106230.

Haywood, L., Kew, F., Bramham, P., Spink, J., Capenhurst, J. and Henry, I. (1989) *Understanding Leisure*. Cheltenham: Stanley Thornes.

Hedrick-Wong, Y. (2007) *The Glittering Silver Marketing: The Rise of Elderly Consumers in Asia*. Singapore: John Wiley and Sons.

Heimtun, B. (2019) Holidays with aging parents: pleasures, duties and constraints. *Annals of Tourism Research*, 76:129–39.

Heinrich, L. and Gullone, E. (2006) The clinical significance of loneliness: a literature review. *Clinical Psychology Review*, 26:695–718.

Heinze, R, Naegele, G and Schneiders, K. (2011) *Wirtschaftliche Potentiale des Alters* (Economic Potential of Old Age). Stuttgart: Kohlhammer.

Helal, A., Mokhtari, M. and Abdulrazak, B. (eds.) (2008) *The Engineering Handbook of Smart Technology for Aging, Disability, and Independence*. New York: Wiley.

Henderson, J. (2007) Population ageing, tourism and travel insurance. *Tourism Recreation Research*, 32(3):79–82.

Henze, M., Alfonso, H., Flicker, L., George, J., Chubb, S. A. P., Hankey, G., Almeida, O., Golledge, J., Norman, P. and Yeap, B. (2017) Profile of diabetes in men aged 79–97 years: the Western Australian Health in Men Study. *Diabetic Medicine*, 34(6):786–93.

Hepple, L. (2001) Multiple regression and spatial policy analysis: George Udny Yule and the origins of statistical social science. *Environment and Planning D: Society and Space*, 19(4):385–407.

Herrmann, M. (2012) Population ageing and economic development: anxieties and policy responses. *Population Ageing* 5(1):23–46.

Hersh, M. (2016) Improving deafblind travelers' experiences: An international survey. *Journal of Travel Research* 55(3):380–94.

Heslop, P., Blaire, P. and Flemming, P. (2013) *Confidential Enquiry into the Premature Deaths of People with Learning Disabilities (CIPOLD)*. London: Norah Fry Research Centre.

Hill, A. (2020) Older people widely demonised in UK. *The Guardian*, 19 March 2020, www.theguardian.com/society/2020/mar/19/older-people-widely-demonised-uk-ageism-report

Hill, C. (2019) Cruise Ship Travel. In J. Keystone, P. Kozarsky, B. Connor, H. Nothdurft, M. Mendelson and K. Leder (eds.), *Travel Medicine* (4th ed.) Oxford: Elsevier, 377–82.

Historic Royal Palaces (2017) *Rethinking Heritage: A Guide to Help Make Your Site More Dementia-Friendly*. www.alzheimers.org.uk/get-involved/dementia-friendly-communities/organisations/dementia-friendly-heritage-sites

Hitt, H. (1954) The role of migration in population change among the aged. *American Sociological Review*, 19:194–200.

HMSO (1954) *Phillips Report: Report by the Committee on the Economic and Financial Problems of the Provision for Old Age*. London: HMSO.

Ho, C.-H. and Peng, H.-H. (2017) Travel motivation for Taiwanese hearing-impaired backpackers. *Asia Pacific Journal of Tourism Research*, 22(4):449–64.

Hobsbawm, E. (2010) *Age of Revolution 1789–1848*. London: Phoenix Press.

Hofäker, D., Hess, M. and König, S. (eds.) (2016) *Delaying Retirement: Progress and Challenges of Active Ageing in Europe, the United States and Japan*. Singapore: Palgrave Macmillan.

Hollenstein, H. (2004) Determinants of the adoption of Information and Communication Technologies (ICT) An empirical analysis based on firm-level data for the Swiss business sector. *Structural Change and Economic Dynamics*, 15(3):315–42.

Holley-Moore, G. and Creighton, H. (2015) *The Future of Transport in an Ageing Society*. London: Age Concern/ILC.

Holmes, K. (2003) Volunteers in the heritage sector: a neglected audience? *International Journal of Heritage Studies*, 9(4): 341–55.

Homans, G. (1958) Social behaviour as exchange. *The American Journal of Sociology*, 63:597–606.

Hong, S., Hasche, L. and Bowland, S. (2009) Structural relationships between social activities and longitudinal trajectories of depression among older adults. *The Gerontologist*, 49:1–11.

Horneman, L., Carter, R., Wei, S. and Ruys, H. (2002) Profiling the senior traveler: an Australian perspective. *Journal of Travel Research*, 41(1), 23–37.

House of Lords (2012) *Ready for Ageing: Report of Sessions 2012–13*. House of Lords: HMSO.

House of Lords (2019) *Select Committee on Intergenerational Fairness and Provision: Tackling Intergenerational Unfairness. Report of Session 2017–19*. HL Paper 329. London: House of Lords.

Hsu, C., Cai, L. and Wong, K. (2007) A model of senior tourism motivations: anecdotes from Beijing and Shanghai. *Tourism Management*, 28(5):1262–73.

Hsu, C. and Kang, S. (2009) Chinese urban mature travelers' motivation and constraints by decision autonomy. *Journal of Travel & Tourism Marketing*, 26(7):703–21.

Huang, L. and Tsai, H. (2003) The study of senior traveler behavior in Taiwan. *Tourism Management*, 24: 561–74.

Huber, D. (2019) A life course perspective to understanding senior tourism patterns and preferences. *International Journal of Tourism Research*, 21(3):372–87.

Hudson, S. (2000) *Snow Business*. London: Continuum.

Hughes, B. and Paterson, K. (2006) The social model of disability and the disappearing body: towards a sociology of impairment. In L. Barton (ed.), *Overcoming Disabling Barriers* (pp. 101–17). London: Routledge.

Hughes, L. (2018) The geriatric 5Ms: An important new construct in geriatric medicine. *GM Journal*, Blog 48 (7/8). www.gmjournal.co.uk/the-geriatric-5ms-an-important-new-construct-in-geriatric-medicine.

Hung, K. and Lu, J. (2016) Active living in later life: an overview of aging studies in hospitality and tourism journals. *International Journal of Hospitality Management*, 53:133–44.

Hunter-Jones, P. (2005) Cancer and tourism. *Annals of Tourism Research*, 32(1):70–92.

Hussain, S., Lei, S., Akram, T., Haider, M., Hussain, S. and Ali, M. (2018) Kurt Lewin's change model: a critical review of the role of leadership and employee involvement in organizational change. *Journal of Innovation and Knowledge*, 3(3):123–7.

Huxtable, R. (2009) The suicide tourist trap: compromise across boundaries. *Journal of Bioethical Inquiry*, 6 (3):327–36.

Hyde, M. and Higgs, P. (2016) *Ageing and Globalisation*. Bristol: Policy Press.

ING (2020) Challenges and opportunities of an ageing society. https://view.ingwb.com/challenges-and-opportunities-of-an-ageing-society.

Inglis, F. (2000) *The Delicious History of the Holiday*. London: Routledge.

Innes, A., Page, S. J. and Cutler, C. (2015) Barriers to leisure participation for people with dementia and their carers: an exploratory analysis of carer and people with dementia's experiences. *Dementia*, 15(6):1643–65.

Institute of Public Policy Research (2009) *The Politics of Ageing*. London: Institute of Public Policy Research.

Iqbal, O., Eklof, B., Tobu, M., and Fareed, J. (2003) Air travel-associated venous thromboembolism. *Medical Principles and Practice*, 12(2):73–80.

Iso-Ahola, S. E. (1980) *Social Psychological Perspectives on Leisure and Recreation*. Springfield, IL: Charles C. Thomas.

Israeli, A. (2002) A preliminary investigation of the importance of site accessibility factors for disabled tourists. *Journal of Travel Research*, 41 (1):101–4.

Iwamasa, G. and Hilliard, K. (1999) Depression and anxiety among Asian American elders: a review of the literature. *Clinical Psychology Review*, 19(3):343–57.

Iwasaki, Y., Coyle, C. and Shank, J. (2010) Leisure as a context for active living, recovery, health, and life quality for persons with mental illness in a global context. *Health Promotion International*, 25(4):483–94.

Iwasaki, Y., Nishino, H., Onda, T. and Bowling, C. (2007) Leisure research in a global world: time or reverse the western domination in leisure research. *Leisure Sciences*, 29:1–5.

Jang, C., Liang, C. and Chen, S. (2019) Spatial dynamic assessment of health risks for urban river cruises. *Environmental Monitoring and Assessment*, 191(1):1.

Jang, J., Blum, A., Liu, J. and Finkel, T. (2018) The role of mitochondria in aging. *The Journal of Clinical Investigation*, 128(9): 3662–70.

Jang, S. and Wu, C. (2006) Seniors' travel motivation and the influential factors: an examination of Taiwanese seniors. *Tourism Management*, 27(2):306–16.

Janke, M., Nimrod, G. and Kleiber, D. (2008) Leisure activity and depressive symptoms of widowed and married women in later life. *Journal of Leisure Research*, 40(2):250–66.

Janoski, T. (1998) *Citizenship and Civil Society*. Cambridge: Cambridge University Press.

Janta, H., Cohen, S. A. & Williams, A. M. (2015). Rethinking visiting friends and relatives mobilities. *Population, Space and Place*, 21(7), 585–98.

Jawad, F. and Kalra, S. (2016) Diabetes and travel. *Journal of the Pakistan Medical Association*, 66(10):1347–8.

Jenkin, C., Eime, R., Westerbeek, H., O'Sullivan, G. and van Uffelen, J. (2017) Sport and ageing: a systematic review of the determinants and trends of participation in sport for older adults. *BMC Public Health*, 17(1): 976.

Jeppsson Grassman, E. and Whitaker, A. (eds.) (2013) *Ageing with Disability: A Lifecourse Perspective*. Bristol: Policy Press.

Johnson, M. (1976) Is 65+ old? *Social Policy* (Nov./Dec.):9–12.

Johnson, J. and Finn, K. (2017) *Designing User Interfaces for an Aging Population*. Cambridge, MA: Morgan Kanfmann.

Johnson, M., Coleman, P. and Bengtson, V. (eds.) (2005) *The Cambridge Handbook on Age and Ageing*. Cambridge: Cambridge University Press.

Jones, D. C. (ed.) (1934) *The Social Survey of Merseyside. Volumes I, II, III*. Liverpool: University Press of Liverpool.

Jones, I., Hyde, M., Victor, C., Wiggins, D., Gilleard, C. and Higgs, P. (2008) *Ageing in a Consumer Society*. Bristol: Policy Press.

Joseph Rowntree Foundation (2004) *From Welfare to Well-Being: Planning for an Ageing Society. Summary conclusions of the Joseph Rowntree Foundation*. Task Group on Housing, Money and Care for Older People. www.jrf.org.uk/sites/default/files/jrf/migrated/files/034.pdf.

Joseph Rowntree Foundation (2017) *UK Poverty 2017*. York: Joseph Rowntree Trust.

Joy, M. (2007) Cardiovascular disease and airline travel. *Heart*, 93(12):1507–09.

Just, R., Hueth, D. and Schmitz, A. (2005) *The Welfare Economics of Public Policy: A Practical Approach to Project and Policy Evaluation*. Cheltenham: Edward Elgar.

Kain, D., Findlater, A., Lightfoot, D., Maxim, T., Kraemer, M., Brady, O., Watts, A., Khan, K. and Bogoch, I. (2019) Factors affecting pre-travel health seeking behaviour and adherence to pre-travel health advice: a systematic review. *Journal of Travel Medicine*, 26(6):taz059.

Kalyani, R., Rodriguez, D., Yeh, H., Golden, S. and Thorpe, R. (2015) Diabetes, race, and functional limitations in older US men and women. *Diabetes Research and Clinical Practice*, 108(3):390–7.

Kanter, M. (2006) Innovation: the classic traps. *Harvard Business Review*, 84(11):1–14.

Kaplan, J. (1946) The psychology of maturity. In P. Harriman (ed.), *The Encyclopaedia of Psychology*. New York: Philosophical Library, 370–8.

Kaplan, M. (1960) *Leisure in America*. New York: John Wiley and Sons.

Karagiannis, N. and King, J. (2019) Introduction: the role of government. In N. Karagiannis, and J. King (eds.), *A Modern Guide to State Intervention: Economic Policies for Growth and Sustainability*. Cheltenham: Edward Elgar Publishing.

Karatepe, O. and ZargarTizabi, L. (2011) Work-related depression in the hotel industry: a study in the United Arab Emirates. *International Journal of Contemporary Hospitality Management*, 2(5):608–23.

Karn, V. (1974) Retiring to the seaside: A study of retirement migration in England and Wales. Unpublished PhD thesis, University of Birmingham.

Karn, V. (1977) *Retiring to the Seaside*. London: Routledge and Kegan Paul.

Kastenholz, E., Eusébio, C. and Figueiredo, E. (2015) Contributions of tourism to social inclusion of persons with disability. *Disability and Society*, 30(8):1259–81.

Kazeminia, A., Del Chiappa, G. and Jafari, J. (2015) Seniors' travel constraints and their coping strategies. *Journal of Travel Research*, 54(1):80–93.

Kelley-Moore, J. (2010) Disability and ageing: the social construction of causality. In D. Dannefer and C. Phillipson (eds.), *The Sage Handbook of Social Gerontology*. London: SAGE, 96–110.

Kelly, J. (ed.) (1993) *Activity and Aging: Staying Involved in Later Life*. Newbury Park, CA: SAGE.

Kim, H., Woo, E. and Uysal, M. (2015) Tourism experience and quality of life among elderly tourists. *Tourism Management*, 46, 465–76.

Kim, J., Kim, J., Williams, R. D., Jr. and Han, A. (2021) The association of social support and leisure time physical activity with mental health among individuals with cancer. *American Journal of Health Promotion*, 35(3):362–8.

Kim, S. and Lehto, X. (2012) The voice of tourists with mobility disabilities: Insights from online customer complaint websites. *International Journal of Contemporary Hospitality Management*, 24(3):451–76.

Kim, S. and Lehto, X. (2013) Leisure travel of families of children with disabilities: Motivation and activities. *Tourism Management*, 37:13–24.

Kim, S., Lee, C. and Klenosky, D. (2003). The influence of push and pull factors at Korean national parks. *Tourism Management*, 24(2): 169–80.

Kim, Y., Weaver, P. and McCleary, K. (1996) A structural equation model: the relationship between travel motivation and information sources in the senior travel market. *Journal of Vacation Marketing*, 3(1): 55–66.

King, R., Warnes, A. M. and Williams, A. M. (2000) *Sunset Lives: British Retirement to Southern Europe*. Oxford: Berg.

Klaver, C., Wolfs, R., Vingerling, J., Hofman, A. and De Jong, P. (1998) Age-specific prevalence and causes of blindness and visual impairment in an older population: The Rotterdam study. *Archives of Ophthalmology*, 116(5):653–8.

Kleiber, D. and Genoe, R. (2012) The relevance of leisure in terms of ageing. In H. Gibson and J. Singleton (eds.), *Leisure and Ageing*. Champaign, IL: Human Kinetics: 43–66.

Klein, E. (2020) Coronavirus will also cause a loneliness epidemic, *Vox*, 12 March 2020. www.vox.com/2020/3/12/21173938/coronavirus-covid-19-social-distancing-elderly-epidemic-isolation-quarantine.

Knight, G. and Bichard, J-A. (2011) *Publicly Accessible Toilets: An Inclusive Design Guide*. London: Royal College of Art.

Ko, H. and Youn, C. (2011) Effects of laughter therapy on depression, cognition and sleep among the community-dwelling elderly. *Geriatrics & Gerontology International*, 11(3):267–74.

Koc, E. (2017) Do all-inclusive holidays promote gluttony, obesity? *The Good Tourism Blog*. https://goodtourismblog.com/2017/06/inclusive-holidays-promote-gluttony-obesity/.

Kocchar, R. and Oates, R. (2014) *Attitudes about Aging: A Global Perspective*. Washington, DC: Pew Research Centre. www.pewresearch.org/

Kocka, J. and Brauer, K. (2010) Civil society and the elderly. In H. Anheier and S. Toepler (eds.), *International Encyclopedia of Civil Society*. New York: Springer.

Komps K. and Johansson, S. (eds.) (2015) *Population Ageing from a Lifecourse Perspective: Critical and International Approaches*. Bristol: Policy Press.

Kong, W. and Loi, K. (2017) The barriers to holiday-taking for visually impaired tourists and their families. *Journal of Hospitality and Tourism Management*, 32:99–107.

Koskella, K. and He, W. (2008) *An Aging World*. Washington, DC: National Institute for Aging and US Bureau of Census.

Koskinen, V. (2019) Spa tourism as a part of ageing well. *International Journal of Spa and Wellness*, 2(1):18–34.

Kotter, J. (1998) *Leading Change*. Boston, MA: Harvard University Press.

Kotter, J. and Cohen, D. (2002) *The Heart of Change: Real-Life Stories of How People Change Their Organizations*. Boston, MA: Harvard Business Review Press.

KPMG (2019a) *Future of Mobility: Travel and Leisure*. https://home.kpmg/uk/en/home/campaigns/2019/09/mobility-2030-future-of-mobility.html

KPMG (2019b) *Global Leisure Perspectives: Leisure Industry Trends from Around the Globe*. https://home.kpmg/uk/en/home/insights/2019/03/global-leisure-perspectives-2019.html

Kreutzwiser, R. (1989) Supply. In G. Wall (ed.), *Outdoor Recreation in Canada*. Toronto: Wiley, 21–41.

Kroesen, M. and Handy, S. (2014) The influence of holiday-taking on affect and contentment. *Annals of Tourism Research*, 45:89–101.

Kubler-Ross, E (1969) *On Death and Dying*. New York: Scribner.

Kumekawa, I. (2017) *The First Serious Optimist: A. C. Pigou and the Birth of Welfare Economics*. Princeton, NJ: Princeton University Press.

Kunreuther, F. (2013) Grassroots associations. In M. Edwards (ed.), *The Oxford Handbook of Civil Society*. Oxford: Oxford University Press, 35–67.

La Plante, M. (2014) Key goals and indicators for successful aging of adults with early-onset disability. *Disability and Health Journal*, 7:S44 – S50.

Lam, K., Chan, C. and Peters, M. (2020) Understanding technological contributions to accessible tourism from the perspective of destination design for visually impaired visitors in Hong Kong. *Journal of Destination Marketing and Management*, 17:100434.

Lane, M. (2007) The visitor journey: the new road to success. *International Journal of Contemporary Hospitality Management*, 19(3):248–54.

Langer, E. (2000) Mindful learning. *Current Directions in Psychological Science*, 9(6):220–3.

Larsen, S., Brun, W. and Øgaard, T. (2009) What tourists worry about: construction of a scale measuring tourist worries. *Tourism Management*, 30 (2):260–5.

Laukkanen, J., Rauramaa, R., Mäkikallio, T., Toriola, A. and Kurl, S. (2011) Intensity of leisure-time physical activity and cancer mortality in men. *British Journal of Sports Medicine*, 45(2):125–9.

Lauría, A. (2016) The Florence experience: a multimedia and multisensory guidebook for cultural towns inspired by Universal Design approach. *Work*, 53(4):709–27.

Lave, J. and Wenger, E. (1991) *Situated Learning: Legitimate Peripheral Participation*. Cambridge: Cambridge University Press.

Law, C. (2002) *Urban Tourism: The Visitor Economy and Growth of Large Citi*es. London: Continuum.

Law, C. and Warnes, A. (1973) The movement of retired people to seaside resorts: A study of Morecambe and Llandudno. *The Town Planning Review*, 44(4):373–90.

Law, C. and Warnes, A. (1976) The changing geography of the elderly in England and Wales. *Transactions of the Institute of British Geographers* New Series 1:453–71.

Lawton, G. and Page, S. J. (1997) Evaluating travel agents' provision of health advice to travellers. *Tourism Management*, 18(2):89–104.

Lawton, M., Moss, M. and Duhamel, L. (1995) The quality of daily life among elderly care receivers. *Journal of Applied Gerontology*, 14(2):50–171.

Layard, R. and Clark, D. (2015) *Thrive: How Better Mental Health Care Transforms Lives and Saves Money*. Harmondsworth: Penguin.

Le Serre, D. and Chevalier, C. (2012) Marketing travel services to senior consumers. *Journal of Consumer Marketing*, 29(4):262–70.

Lee, B., Agarwal, S. and Kim, H. (2012) Influences of travel constraints on the people with disabilities' intention to travel: An application of Seligman's helplessness theory. *Tourism Management*, 33(3):569–79.

Lee, C. and King, B. (2019) Determinants of attractiveness for a seniors-friendly destination: A hierarchical approach. *Current Issues in Tourism*, 22(1):71–90.

Lee, H-Y., Yu, C-P., Wu, C-D. and Pan, W-C. (2018) The effects of leisure activity diversity and exercise time on the prevention of depression in the middle aged and elderly residents of Taiwan. *International Journal of Environmental Health Research and Public Health* 15(4):654.

Lee, J., Lau, S., Meijer, E. and Hu, P. (2020) Living longer, with or without disability? A global and longitudinal perspective. *Journals of Gerontology – Series A Biological Sciences and Medical Sciences*, 75(1), 162–7.

Lee, R. and Mason, A. (2010) Some macroeconomic aspects of global population aging. *Demography*, 47(1):151–72.

Lefebvre, H. (1968) *Le Droit à la Ville*. Paris: Anthropos.

Lefebvre, H. (1973) La survie du capitalisme; la re-production des rapports de production. Trans. Frank Bryant as The Survival of Capitalism. London: Allison and Busby, 1976.

Lefebvre, H. (1991) *The Production of Space*. Trans. by Donald Nicholson-Smith. Oxford: Basil Blackwell.

Leggat, P. (2005) Travel medicine: an Australian perspective. *Travel Medicine and Infectious Disease*, 3(2):67–75.

Lehto, X., Jang, S., Achana, F. and O'Leary, J. (2008) Exploring tourism experience sought: a cohort comparison of Baby Boomers and the Silent Generation. *Journal of Vacation Marketing*, 14(3):237–52.

Leitner, M. and Leitner, S. (1995) *Leisure in Later Life.* (2nd edition). New York: Haworth Press.

Lemon, B., Bengtson, V. and Peterson, J. (1972) An exploration of the activity theory of aging: activity types and life satisfaction among in-movers to a retirement community. *Journal of Gerontology*, 27(4):511–23.

Leonardi, M., Bickenbacj, J. Ustun, T., Kostanjsek, N. and Chatterjei. S. (2006) The definition of disability: what's in a name? *The Lancet*, 368(9543):1219–21.

Levi, E., Dolev, T., Colins-Kreiner, N. and Zilcha-Mano, S. (2019) Tourism and depressive symptoms. *Annals of Tourism Research*, 74:191–4.

Lewin, K. (1947) *Field Theory in Social Science.* New York: Harper Row.

Lewis, M. and Butler, R. (1972) Why is women's lib ignoring old women. *International Journal of Aging: Human Development*, 3:223–31.

Li, J., Loerbroks, A. and Angerer, P. (2013) Physical activity and risk of cardiovascular disease: what does the new epidemiological evidence show? *Current Opinion in Cardiology*, 28(5):575–83.

Lieux, E., Weaver, P. and McCleary, K. (1994) Lodging preferences of the senior tourism market. *Annals of Tourism Research*, 21(4):712–28.

Linder, S. (1969) *The Harried Leisure Class.* New York: Colombia University Press.

Lindqvist, L. and Bjork, P. (2000) Perceived safety as an important quality dimension among senior tourists. *Tourism Economics*, 6(2):151–8.

Little, V. (2008) An overview of research using the time-budget methodology to study age-related behaviour. *Ageing and Society*, 4(1):3–20.

Littrell, M., Paige, R. and Song, K. (2004) Senior travellers: tourism activities and shopping behaviours. *Journal of Vacation Marketing*, 10(4):348–62.

Liu, Y. (2009) Sport and social inclusion: evidence from the performance of public leisure facilities. *Social Indicators Research*, 90:325–37.

Lloyd, J. and Lord, C. (2015) *Defined Capability: Pensions, Financial Capability and decision-making among Retirees.* London: Strategic Society Centre.

Local Government Association. (2012) *Dementia Friendly Communities: Guidance for Councils.* London: Local Government Association.

Loch, L. and French, J. (1948) Overcoming resistance to change. *Human Relations*, 1(4):512–32.

Löfgren, O. (2002) *On Holiday: A History of Vacationing.* Berkeley and Los Angeles: California University Press.

Lokon, E., Sauer, P. and Li, Y. (2016) Activities in dementia care: A comparative assessment of activity types. *Dementia*, 18(2):471–89.

Lomas, A., Leonardi-Bee, J. and Bath-Hextall, F. (2012) A systematic review of worldwide incidence of nonmelanoma skin cancer. *British Journal of Dermatology*, 166(5):1069–80.

Longino, C. and Kart, C. (1982) Explicating activity theory: A formal replication. *Journal of Gerontology*, 37(6):713–22.

Loos, E. Sourbati, M. and Behrendt, F. (2016) The role of mobility digital ecosystems for the elderly. *TEME Journal of Land Use, Mobility and Environment*, 17:7465.

Lord, S. and Close, J. (2018) New horizons in falls prevention. *Age and Ageing*, 47(4):492–8.

Losada, N., Alén, E., Cotos-Yáñez, T. and Domínguez, T. (2019) Spatial heterogeneity in Spain for senior travel behavior. *Tourism Management*, 70:444–52.

Lovelock, C., Patterson, P. and Walker, R. (2001) *Services Marketing: An Asia-Pacific Perspective*. French's Forest, NSW: Prentice-Hall.

Low, J. and Chan, D. (2002) Air travel in older people. *Age and Ageing*, 31(1):17–22.

Lu, J., Hung, K., Wang, L., Schuett, M., and Hu, L. (2016) Do perceptions of time affect outbound-travel motivations and intention? An investigation among Chinese seniors. *Tourism Management*, 53: 1–12.

Lu, L. and Hu, C. (2005) Personality, leisure experiences and happiness. *Journal of Happiness Studies*, 6:325–42.

Luiu, C., Tight, M. and Burrow, M. (2018) An investigation into the factors influencing travel needs during later life. *Journal of Transport & Health*, 11:86–99.

Lundberg, G., Komarovsky, M. and McInerny, M. (1934) *Leisure: A Suburban Study*. New York: Columbia University Press.

Lye, M. and Donnellan, C. (2000). Heart disease in the elderly. *Heart*, 84(5):560.

Lyu, S. (2017) Which accessible travel products are people with disabilities willing to pay more? A choice experiment. *Tourism Management*, 59:404–12.

Mack, J. and Lansley, S. (1985) *Poor Britain*. London: George Allen & Unwin.

Mackey, J. and Sisodia, R. (2013) *Conscious Capitalism: Liberating the Heroic Spirit of Business*. Boston, MA: Harvard Business School Publishing Corporation.

Macpherson, J. (1869) *The Baths and Wells of Europe: Their Actions and Uses*. London: Macmillan.

Maddox, G. (1968) Persistence of lifestyle among the elderly: a longitudinal study of patterns of social activity in relation to life satisfaction. In B. Neugarten (ed.), *Middle Age and Aging: A Reader in Social Psychology*. Chicago: Chicago University Press, 181–83.

Mahesh, K. and Suresh, J. (2009) Knowledge criteria for organization design. *Journal of Knowledge Management*, 13(4):41–51.

Mak, J., Carlile, L. and Dai, S. (2005) Impact of population aging on Japanese international travel to 2025. *Journal of Travel Research*, 44(2):151–62.

Makoni, S. (2008) Aging in Africa: a critical review. *Journal of Cross-Cultural Gerontology*, 23(2):199–209.

Makoni, S. and Stroeken, K. (eds.) (2017) *Ageing in Africa: Sociolinguistic and Anthropological Approaches*. London: Routledge.

Manton, K. (1982) Changing concepts of morbidity and mortality in the elderly population. *Millbank Memorial Fund Quarterly* 60: 183–244.

Marc, M. and Meyersohn, E. (eds.) (1955) *Mass Leisure*. New York: Free Press.

Marcolongo, T., Valk, T. and Jones, M. (2019) Mind the gap: building the psychological capital of travellers. *Journal of Travel Medicine*, 26(1):tay142.

Marshall, L. (2006) Aging: A feminist issue. *NWSA Journal* 18(1): vii–xiii.

Marshall, T. (1964) *Class, Citizenship and Social Development*. Garden City, NY: Doubleday and Company.

Massingham, P. (2020) *Knowledge Management: Theory in Practice*. London: SAGE.

McCabe, S and Johnson, S. (2013) The happiness factor in tourism: subjective well-being and social tourism. *Annals of Tourism Research*, 41(1):42–65.

McCabe, S., Minnaert, L. and Diekmann, A. (2011) *Social Tourism in Europe: Theory and Practice*. Bristol: Channel View Publications.

McCallum, J. (1988). Japanese Teinen Taishoku: how cultural values affect retirement. *Ageing and Society*, 8(1):23–41.

McDermott, R. (1999) Why information technology inspired but cannot deliver knowledge management. *California Management Review*, 41(4): 103–17.

McGuire, F., Boyd, R., Janke, M. and Aybar-Darnell, B. (2013) *Leisure and Aging: Ulyssean Living in Later Life* (5th ed.). Champaign, IL: Sagamore.

McHugh, K., Hogan, T. and Happel, S. (1995) Multiple residence and cyclical migration: a life course perspective. *The Professional Geographer*, 47(3):251–67.

McHugh, K. and Mings, R. (1991) On the road again: seasonal migration to a sunbelt metropolis. *Urban Geography*, 12(1):1–18.

McIntosh, A. (2020) The hidden side of travel: epilepsy and tourism. *Annals of Tourism Research*, 81:102856.

McIntosh, I. (1998) Health hazards and the elderly traveler. *Journal of Travel Medicine*, 5(1):27–9.

McKercher, B. (2018) The impact of distance on tourism: a tourism geography law. *Tourism Geographies*, 20 (5):905–9.

McKercher, B. and Darcy, S. (2018) Re-conceptualizing barriers to travel by people with disabilities. *Tourism Management Perspectives*, 26:59–66.

McLachlin, L. and Claflin, T. (2004) Recreation for special populations: an overview. In M. Leitner and S. Leitner (eds.), *Leisure Enhancement* (3rd edition). Binghampton, NJ: Haworth Press, 341–62.

McLaughlin, K., Osborne, S. and Ferlie, E. (eds.) (2002) *New Public Management: Current Trends and Future Prospects*. London: Psychology Press.

McLeish, J. (1976) *The Ulyssean Adult: Creativity in Middle and Later Years*. Toronto: McGraw-Hill.

McPherson, G., Oluwaseyi, A., McGillivray and Misener, L. (2020) Disability and events. In S. J. Page and J. Connell (eds.), *The Routledge Handbook of Events* (2nd edition). London: Routledge, 491–501.

Meiners, N. (2014) Economics of ageing: research area and perspectives. *Quality in Ageing and Older Adults*, 15(2):63–75.

Mellor, H. (1962) Retirement to the coast. *The Town Planning Review*, 33(1):40–8.

Menec, V. (2003) The relation between everyday activities and successful aging: a 6-year longitudinal study. *The Journals of Gerontology: Series B*, 58(2):S74 – S82.

Mercer (2019) *Melbourne Mercer Global Pension Index*. Melbourne: Monash Centre for Financial Studies.

Mercer, D. (1970) The geography of leisure – A contemporary growth-point. *Geography*, 55(3):261–73.

Merriam, S. (2009) *Qualitative Research: A Guide to Design and Implementation*. San Francisco: Wiley.

Mesquita, S. and Carneiro, M. (2016) Accessibility of European museums to visitors with visual impairments. *Disability and Society*, 31(3):373–88.

Mezrad, R. (ed.) (2020) *Obesity: Global Impact and Epidemiology*. Oxford: Elsevier.

Mihalič, T. (2003) Supply. In J. Jenkins and J. Pigram (eds.), *Encyclopedia of Leisure and Outdoor Recreation*. Routledge: London, 489–92.

Miles, W. (1942) Psychological aspects of ageing. In E. Cowdry (ed.), *Problems of Ageing: Biological and Medical Aspects* (2nd edition). The Williams and Wilkins Company, Baltimore, MD, 756–74.

Milligan, C., Gatrell, A. and Bingley, A. (2004) 'Cultivating health': Therapeutic landscapes and older people in northern England. *Social Science and Medicine* 58: 1781–93.

Milman, A. (1998) The impact of tourism and travel experience on senior travelers' psychological well-being. *Journal of Travel Research*, 37(2):166–70.

Milne, A. (2010) The 'D' word: reflections on the relationship between stigma, discrimination and dementia. *Journal of Mental Health*, 19(3):227–33.

Minkler, M. and Fadem, P. (2002) 'Successful aging': a disability perspective. *Journal of Disability Policy Studies*, 12(4):229–35.

Minnaert, L., Maitland, R. and Miller, G. (2011) What is social tourism? *Current Issues in Tourism*, 14(5):403–15.

Mitas, O., Yarnal, C. and Chick, G. (2012) Jokes build community: mature tourists' positive emotions. *Annals of Tourism Research*, 39(4):1884–1905.

Mizumoto, K., Kagaya, Katsushi, Z. and Alexander, C. (2020) Estimating the asymptomatic proportion of coronavirus disease 2019 (COVID-19) cases on board the Diamond Princess cruise ship, Yokohama, Japan, 2020. *Euro Surveillance*, 25(10):pii=2000180.

Moal-Ulvoas, G. (2017) Positive emotions and spirituality in older travelers. *Annals of Tourism Research*, 66:151–8.

Mollenkopf, H., Hieber, A. and Wahl, H. (2011) Continuity and change in older adults' perceptions of out-of-home mobility over ten years: a qualitative – quantitative approach. *Ageing and Society*, 31(5), 782–802.

Möller, C., Weiermair, K. and Wintersberger, E. (2007) The changing travel behaviour of Austria's ageing population and its impact on tourism. *Tourism Review*, 62(3–4):15–20.

Molton, I. and Ordway, A. (2019) Aging with disability: populations, programs, and the new paradigm. An introduction to the special issue. *Journal of Aging and Health*, 31 (10 supplement):3S – 20S.

Moore, B. and Van Nierop, H. (eds.) (2006) *Twentieth Century Mass Society in Britain and the Netherlands*. London: Bloomsbury.

Morad, T. (2007) Tourism and disability: a review of cost-effectiveness. *International Journal on Disability and Human Development*, 6(3):279–82.

Morgan, N., Pritchard, A. and Sedgley, D. (2015) Social tourism and well-being in later life. *Annals of Tourism Research*, 52:1–15.

Morrison, A. (2019) *Marketing and Managing Tourism Destinations* (2nd edition). London: Routledge.

Moschis, G. (1996) *Gerontographics: Lifestage Segmentation for marketing Strategy Development*. Westport, CT: Quorun.

Moschis, G. and Ünal, B. (2008) Travel and leisure services preferences and patronage motives of older consumers. *Journal of Travel and Tourism Marketing*, 24(4):259–69.

Moulaert, F., MacCallum, D., Mehmood, A. and Hamdouch, A. (eds.) (2013) *The International Handbook on Social Innovation: Collective Action, Social Learning and Transdisciplinary Research*. Cheltenham: Edward Elgar Publishing.

Moura, A. Kastenholz, E. and Pereira, A. (2018) Accessible tourism and its benefits for coping with stress. *Journal of Policy Research in Tourism, Leisure and Events*, 10(3):241–64.

Muller, T. and O'Cass, A. (2001) Targeting the young at heart: seeing senior vacationers the way they see themselves. *Journal of Vacation Marketing*, 7(4): 285–301.

Murthy, V. (2020) *Together: Loneliness, Health and What Happens When We Find Connection*. London: Pinto Books in association with the Wellcome Collection.

Musa, G. and Sim, O. (2010) Travel behaviour: a study of older Malaysians. *Current Issues in Tourism*, 13(2):177–92.

Musselwhite, C., Holland, C. and Walker, I. (2015) The role of transport and mobility in the health of older people. *Journal of Transport & Health*, 2(1):1–4.

Nagi, S. (1969) *Disability and Rehabilitation: Legal Clinical and Self-Concepts and Measurement*. Columbus, OH: Ohio State University Press.

Nagi, S. (1977) The disabled and rehabilitation services: a national overview. *American Rehabilitation*, 2(5):26–33.

Nagi, S. (1991) Disability concepts revisited: implications for prevention. In A. Pope and A. Tarlow (eds.), *Disability in America: Towards a National Agenda for Prevention.* Washington, DC: National Academic Press, 309–27.

Nair, K. (2005) The physically ageing body and the use of space. In G. Andrews and D. Phillips (eds.), *Ageing and Place: Perspectives, Policy, Practice.* London: Routledge, 110–17.

National Autistic Society and Visit England (2018) *Welcoming Autistic People: A Guide for Tourism Venues.* www.visitbritain.org/sites/default/files/vb-corporate/business-hub/resources/autism_guide_for_tourism_venues_2018.pdf.

Nawijn, J. (2016) Positive psychology in tourism: a critique. *Annals of Tourism Research,* 56:151–3.

Nazari Adli, S. and Donovan, S. (2018) Right to the city: applying justice tests to public transport investments. *Transport Policy,* 66:56–65.

Newman, A. B., Kritchevsky, S. B., Guralnik, J. M., Cummings, S. R., Salive, M., Kuchel, G. A., … Ferrucci, L. (2020) Accelerating the search for interventions aimed at expanding the health span in humans: the role of epidemiology. *Journals of Gerontology – Series A Biological Sciences and Medical Sciences,* 75(1), 77–86.

Newton, J. (2018) Too fat for that ass: Greece bans obese tourists from riding donkeys after animal rights activists shed light on abuse. *Daily Mail,* 9 October.

NHS Scotland (n.d.) Mental health and travel. www.fitfortravel.nhs.uk/advice/general-travel-health-advice/mental-health-and-travel#risk

Nielsen, K. (2014) Approaches to seniors' tourist behaviour. *Tourism Review,* 69(2):111–21.

Nilsen, P. (2015) Making sense of implementation theories, models and frameworks. *Implementation Science,* 10(1):53.

Nilsson, I., Nyqvist, F., Gustafson, Y. and Nygard, M. (2015) Leisure engagement: medical conditions, mobility difficulties, and activity limitations – a later life perspective. *Journal of Aging Research:* 610154.

Nimrod, G. (2008) Retirement and tourism: themes in retirees' narratives. *Annals of Tourism Research,* 35(4):859–78.

Nimrod, G. (2010) Seniors' online communities: a quantitative content analysis. *The Gerontologist,* 50(3):382–92.

Nimrod, G. (2014) The benefits of and constraints to participation in seniors' online communities. *Leisure Studies,* 33(3):247–66.

Nimrod, G. and Adoni, H. (2006) Leisure-styles and life satisfaction among recent retirees in Israel. *Ageing and Society,* 26(4):607–30.

Nimrod, G. and Kleiber, D. (2007) Reconsidering continuity and change in later life: toward an innovation theory of successful aging. *International Journal of Aging and Human Development,* 65.1:2–22.

Nimrod, G., Kleiber, D. and Berdychevsky, L. (2012) Leisure in coping with depression. *Journal of Leisure Research,* 44(4):419–49.

Nimrod, G. and Rotem, A. (2011) An exploration of the innovation theory of successful ageing among older tourists. *Ageing and Society,* 32(3):379–404.

Nimrod, G. and Shrira, A. (2016) The paradox of leisure in later life. *The Journals of Gerontology:Series B,* 71(1):106–11.

Nordbakke, S. and Schwanen, T. (2014) Well-being and mobility: a theoretical framework and literature review focusing on older people. *Mobilities,* 9(1):104–29.

Norman, W., Daniels, M., McGuire, F. and Norman, C. (2001) Whither the mature market: an empirical examination of the travel motivations of neo-mature and veteran-mature markets. *Journal of Hospitality and Leisure Marketing,* 8(3–4):113–30.

Northcott, H. and Petruik, C. (2011) The geographic mobility of elderly Canadians. *Canadian Journal on Aging*, 30(3):311–22.

Nyce, S. A. and Schieber, S. J. (2005) *The Economic Implications of Aging Societies – The Costs of Living Happily Ever After*. Cambridge: Cambridge University Press.

Nyman, E., Westin, K. and Carson, D. (2018) Tourism destination choice sets for families with wheelchair-bound children. *Tourism Recreation Research*, 43(1):26–38.

O'Connell, J. (2017) Feeling lonely? You're far from alone. *Irish Times*, 7 January. www.irishtimes.com/life-and-style/people/feeling-lonely-you-re-far-from-alone-1.2924443

OECD (2018) *OECD Trends and Policies 2018*. Paris: OECD.

OECD (2019) *Pensions at a Glance. OECD and G20 Indicators*. Paris: OECD.

Office of the First and Deputy Minister (2004) *Ageing in an Inclusive Society*. Belfast: Office of the First and Deputy Minister.

Oldenburg, M., Herzog, J., Püschel, K. and Harth, V. (2016) Mortality of German travellers on passenger vessels. *Journal of Travel Medicine*, 23(1):1–5.

Oliver, M. (1992) Changing the social relations of research production? *Disability, Handicap and Society*, 7(2):101–20.

Oliver, M. (1990) *The Politics of Disablement*. Basingstoke: Macmillan.

Oliver, M. (1996) *Understanding Disability: From Theory to Praxis*. Basingstoke: Palgrave.

Oliver, M. (2013) The social model of disability: Thirty years on. *Disability & Society*, 28(7):1024–6.

ONS (2015) *Leisure Time in the UK*. London: ONS.

ONS (2019) *Annual Survey of Hours and Earnings*. www.ons.gov.uk/searchdata?q=Annual Survey of Hours and Earnings

Osburn, J., Caruso, G. and Wolfensberger, W. (2011) The concept of 'Best Practice': a brief overview of its meanings, scope, uses, and shortcomings. *International Journal of Disability, Development and Education*, 58(3):213–22.

Osgood, N. (ed.) (1982) *Life After Work: Retirement, Leisure, Recreation and the Elderly*. New York: Praeger.

Otoo, F. and Kim, S. (2020) Analysis of studies on the travel motivations of senior tourists from 1980 to 2017: progress and future directions. *Current Issues in Tourism*, 23 (4):393–417.

Ozturk, Y., Yayli, A. and Yesiltas, M. (2008) Is the Turkish tourism industry ready for a disabled customer's market? *Tourism Management*, 29(2):382–9.

Packer, T., Small, J. and Darcy, S. (2008) *Tourist Experiences of Visitors With Visual Impairment*. Southport, QLD: Sustainable Tourism Cooperative Research Centre. https://opus.lib.uts.edu.au/handle/10453/12248

Paddison, B. and Walmesley, A. (2018) New Public Management in tourism: a case study of York. *Journal of Sustainable Tourism*, 26(6):910–26.

Pagán, R. (2012) Time allocation in tourism for people with disabilities. *Annals of Tourism Research*, 39(3):1514–37.

Pagán, R. (2015) The contribution of holiday trips to life satisfaction: the case of people with disabilities. *Current Issues in Tourism*, 18(6):524–38.

Pagán, R. (2020) How important are holiday trips in preventing loneliness? Evidence for people without and with self-reported moderate and severe disabilities. *Current Issues in Tourism*, 23(11):1394–1406.

Paganini-Hill, A. (2011) Lifestyle practices and cardiovascular disease mortality in the elderly: the leisure world cohort study. *Cardiology Research and Practice*, 983764.

Page, S. J. (2019) *Tourism Management: An Introduction* (5th ed.). London: Routledge.

Page, S. J. (2021) 'Foreword'. In M. Ferrante, O. Fritz and Ö. Öner (eds.), *Regional Science Perspectives on Tourism and Hospitality*. Bern: Springer Nature.

Page, S. J. and Connell, J. (2010) *Leisure: An Introduction*. Harlow: Pearson Education.

Page, S. J. and Connell, J. (2020) *Tourism: A Modern Synthesis* (5th ed.). London: Routledge.

Page, S. J., Bentley, T. and Walker, L. (2005) Scoping the nature and extent of adventure tourism operations in Scotland: how safe are they? *Tourism Management*, 26(3):381–97.

Page, S. J., Bentley, T., Meyer and Chalmers, D. (2001) Scoping the extent of tourist road safety: Motor vehicle transport accidents in New Zealand 1982–1996. *Current Issues in Tourism*, 4(6): 503–26.

Page, S. J., Innes, A. and Cutler, C. (2015) Developing Dementia-Friendly tourism destinations: an exploratory analysis. *Journal of Travel Research*, 54(4):467–81.

Page, S. J. and Meyer, D. (1996) Tourist accidents: an exploratory analysis. *Annals of Tourism Research*, 23(3):666–90.

Page, S. J., Yeoman, I., Connell, J. and Greenwood, C. (2010) Scenario planning as a tool to understand uncertainty in tourism: the example of transport and tourism in Scotland in 2025. *Current Issues in Tourism*, 13(2):99–137.

Paixao, M., Dewar, R., Cossar, J. and Reid, D. (1991) What do Scots die of when abroad? *Scottish Medical Journal*, 3(4):114–16.

Pak, T. (2019) Old-age income security and tourism demand: a quasi-experimental study. *Journal of Travel Research*, 59(7):1298–1315.

Palmer, S., Albergante, L. and Blackburn, C. and Newman, T. J. (2018) Thymic involution and rising disease incidence with age. *Proceedings of the National Academy of Sciences*, 115(8):1883–8.

Parker, S. (1976) *The Sociology of Leisure*. London: George Allen and Unwin.

Patmore, J. A. (1983) *Recreation and Resources*. Oxford: Blackwell.

Patterson, I. (2018) *Tourism and Leisure Behaviour in an Ageing World*. Wallingford: CABI.

Patterson, I. and Pegg, S. (2009a) Marketing the leisure experience to baby boomers and older tourists. *Journal of Hospitality Marketing & Management*, 18(2–3):254–72.

Patterson, I. and Pegg, S. (2009b) Serious leisure and people with intellectual disabilities: benefits and opportunities. *Leisure Studies*, 28(4):387–402.

Patuelli, R. and Nijkamp, P. (2016) Travel motivations of seniors: A review and a meta-analytical assessment. *Tourism Economics*, 22(4):847–62.

Peake, D., Gray, C., Ludwig, M. and Hill, C. (1999) Descriptive epidemiology of injury and illness among cruise ship passengers. *Annals of Emergency Medicine*, 33(1):67–72.

Pearce, D. (2015) Destination management in New Zealand: Structures and functions. *Journal of Destination Marketing and Management*, 4(1):1–12.

Pearce, P. (1993) Fundamentals of tourist motivation. In D. Pearce and R. Butler (eds.), *Tourism Research: Critiques and Challenges*. London: Routledge.

Pearce, P. and Singh, S. (1999) Senior tourism. *Tourism Recreation Research*, 24(1):1–4.

Pearce, S., Kellaher, L. and Holland, C. (2006) *Environment and Identity in Later Life*. Maidenhead: Open University.

Peng, H., Ho, C. and Chan, D. (2013) Hands-eyes versus mouth-ears: exploring the consumer value of hearing-impaired tourists in an outbound group package tour. *Journal of Outdoor Recreation Studies*, 26(2):103–27.

Pensions Watch (2020) Pensions Watch Database. www.pension-watch.net/

Perlman, D. and Peplau, L. A. (1981) Towards a social psychology of loneliness. In R. Gilmour and S. Duck (eds.), *Personal Relationships: 3, Relationships in Disorder*. London: Academic Press, 31–56.

Pernecky, T. (2016) *Approaches and Methods in Event Studies*. London: Routledge.

Pettigrew S. (2007) Reducing the experience of loneliness among older consumers. *Journal of Research for Consumers*, 12:1–4.

Pfeiffer, E. (ed.) (1974) *Successful Aging: A Conference Report*. Durham, NC: Duke University.

Phillips, P., Page, S. J. and Sebu, J. (2020a) Achieving research impact in tourism: modelling and evaluating outcomes from the UKs Research Excellence Framework. *Tourism Management*, 78:104072.

Phillips, P., Page, S. J. and Sebu, J. (2020b) Business and management research themes and impact. *Emerald Open Research*, 2:67 https://doi.org/10.35241/emeraldopenres.13987.1

Phillipson, C., Bernard, M., Phillips, J. and Ogg, J. (2001) *The Family and Community Life of Older People: Social Networks and Social Support in three Urban Areas*. London: Routledge.

Pigou, A. (1920) *The Economics of Welfare*. London: Macmillan.

Pike, S. and Page, S. J. (2014) Destination Marketing Organizations and destination marketing: a narrative analysis of the literature. *Tourism Management*, 41:202–27.

Piramanayagam, S., Seal, P. and More, B. (2019) Inclusive hotel design in India: a user perspective. *Journal of Accessibility and Design for All*, 9(1):41–65.

Pisutsan, P., Soonthornworasiri, N., Matsee, W., Phumratanaprapin, W., Punrin, S., Leowattana, W., … Piyaphanee, W. (2019) Incidence of health problems in travelers to Southeast Asia: A prospective cohort study. *Journal of Travel Medicine*, 26(7):taz045. doi:10.1093/jtm/taz045

Plouffe, L. and Kalache, A. (2011) Making communities age friendly: state and municipal initiatives in Canada and other countries. *Gaceta Sanitaria*, 25:131–37.

Plush, H. (2017) What it's really like to travel as a blind person. *The Telegraph*, 3 February. www.telegraph.co.uk/travel/travel-truths/what-its-like-to-travel-as-a-blind-person/

Pochun, M. (1999) Ageing of population in the Republic of Mauritius: implications for senior tourism. *Tourism Recreation Research*, 24(1):93–5.

Pol, E. and Ville, S. (2009) Social innovation: buzz word or enduring term? *The Journal of Socio-Economics*, 38(6): 878–85.

Polanyi, M. (1967) *The Tacit Dimension*. New York: Anchor.

Poon, A. (1993) *Tourism, Technology and Competitive Strategies*. Wallingford: CABI.

Pope, A. and Tarlow, A. (1991) *Disability in America: Towards a National Agenda for Prevention*. Washington: National Academy Press.

Popple, K. and Redmond, M. (2000) Community development and the voluntary sector in the new millennium: the implications of the Third Way in the UK. *Community Development Journal*, 35(4):391–400.

Poria, Y., Reichel, A. and Brandt, Y. (2009) People with disabilities visit art museums: an exploratory study of obstacles and difficulties. *Journal of Heritage Tourism*, 4(2):117–29.

Poria, Y., Reichel, A., and Brandt, Y. (2011) Blind people's tourism experiences: an exploratory study. In D. Buhalis and S. Darcy (eds.), *Accessible Tourism: Concepts and Issues*. Bristol: Channel View Publications, 149–59.

Porter, M. (1985) *Competitive Advantage: Creating and Sustaining Superior Performance*. New York: The Free Press.

Porter, M. and Kramer, M. (2011) Creating shared value. *Harvard Business Review*, Jan. – Feb., 1–17.

Possick, S. and Barry, M. (2006) Air travel and cardiovascular disease. *Journal of Travel Medicine*, 11(4):243–50.

Povey, C., Mills, R., and Gomez de la Cuesta, G. (2011) Midlife and beyond – Autism and ageing: Issues for the future. *Geriatric Medicine*, April:230–2.

Powell, J. (2006) *Social theory and ageing*. Lanham, MD: Rowman & Littlefield Publishers Inc.

Pratt, S., Tolkach, D. and Kirillova, K. (2019) Tourism and death. *Annals of Tourism Research*, 78:102758.

Prayag, G. (2012) Senior travelers' motivations and future behavioral intentions: the case of NICE. *Journal of Travel & Tourism Marketing*, 29(7):665–81.

Prideaux, B., Wei, S. and Ruys, H. (2001) The senior drive tour market in Australia. *Journal of Vacation Marketing*, 7(3):209–19.

Prosci (n.d.) The Prosci ADKAR model. www.prosci.com/methodology/adkar

Public Health Agency of Canada (2012) Age-Friendly Communities in Canada: Community Implementation Guide. www.canada.ca/en/public-health/services/publications/healthy-living/age-friendly-communities-canada-community-implementation-guide.html

Putnam, R. (2000) *Bowling Alone: America's Declining Social Capital*. New York: Palgrave Macmillan.

Quadango, J. (2007) *Ageing and the Life Course: An Introduction to Social Gerontology* (4th edition). McGraw-Hill.

Quetelet, A. (1836) *Sur l'homme et la développement de ses facultés, ou essai de physique sociale* (On Man and the Development of His Faculties, or Essay of Social Physics). Paris: Bachelier, Imprimeur-Libraire.

Ragsdell, G. (2016) Knowledge management in the not-for-profit sector. *Journal of Knowledge Management*, 20(1).

Randle, M. and Dolnicar, S. (2019) Enabling people with impairments to use Airbnb. *Annals of Tourism Research*, 76:278–89.

Rawls, J. (1971) *A Theory of Justice*. Cambridge, MA: Belknap Press.

Ray, N. and Ryder, M. (2003) 'Ebilities' tourism: an exploratory discussion of the travel needs and motivations of the mobility-disabled. *Tourism Management*, 24(1):57–72.

Ray, R. (1996) A postmodern perspective on feminist gerontology. *The Gerontologist*, 36:674–80.

Reid, C. (2017) The Global Epidemiology of Tourist Fatalities. (Masters of Education, Human Movement, Sport and Leisure, Bowling Green State University.)

Rhoden, S., Ineson, E. and Ralston, R. (2009) Volunteer motivation in heritage railways: a study of the West Somerset Railway volunteers. *Journal of Heritage Tourism*, 4(1): 19–36.

Richards, V., Pritchard, A. and Morgan, N. (2010) (Re)Envisioning tourism and visual impairment. *Annals of Tourism Research*, 37(4): 1097–1116.

Ritchie, J. and Lewis, J. (eds.) (2003) *Qualitative Research Practice: A Guide for Social Science Students and Researchers*. London: SAGE.

Riva, G., Marsan, A. and Grassi, C. (eds.) (2014) *Active Ageing and Healthy Living*. Amsterdam: IOS Press.

Roadburg, A. (1985) *Aging, Retirement, Leisure and Work in Canada*. Toronto: Methuen.

Robert, L. (2006) An original approach to aging: An appreciation of Fritz Verzár's contribution in the light of the last 50 years of gerontological facts and thinking. *Gerontology*, 52(5), 268–74.

Robert, L. and Labat-Robert, J. (2017) Comments on the history of medical – biological studies of aging, the birth of scientific gerontology. *Current Research in Translational Medicine*, 65(1), 44–7.

Roberts, K. (2004) *The Leisure Industries*. Palgrave: Basingstoke.

Roebuck J. (1979) When does old age begin? The evolution of the English definition. *Journal of Social History*, 12(3):416–28.

Rogers, E. (1962) *Diffusion of Innovations*. New York: The Free Press.

Rogers, N., Hawkins, B. and Eklund, S. (1998) The nature of leisure in the lives of older adults with intellectual disability. *Journal of Intellectual Disability Research*, 42(2):122–30.

Rogers, T. (1974) Migration of the aged population. *International Migration*, 12:61–70.

Roland, K. and Chappell, N. (2015) Meaningful activity for persons with dementia: family caregiver perspectives. *American Journal of Alzheimer's Disease & Other Dementias*, 30(6):559–68.

Romsa, G. and Blenman, M. (1989) Vacation patterns of the elderly German. *Annals of Tourism Research*, 16(2):178–88.

Rosenteil, T. (2007) A nation of 'haves' and 'have nots'? www.pewresearch.org/

Ross, G. (2005) Senior tourists sociability and travel preparation. *Tourism Review*, 60(2):6–15.

Rotherham, I. (2014) *Spas and Spa Visiting*. London: Bloomsbury Publishing.

Roulstone, A., Thomas, C. and Watson, N. (2012) The changing terrain of disability studies. In N. Watson, A. Roulstone and C. Thomas (eds.), *The Routledge Handbook of Disability Studies*. London: Routledge, 3–11.

Rowe, J. and Kahn, R. (1987) Human aging: usual and successful. *Science* 237:143–9.

Rowles, G. (1978) *Prisoners of Space?* Boulder, CO: Westview.

Rowles, G. (1986) The geography of ageing and the aged: toward an integrated perspective. *Progress in Human Geography*, 10(4):511–39.

Rowntree, S. (1901) *Poverty: A Study of Town Life*. London: Macmillan and Co.

Rowntree, S. (1947) *Old People: Report of a Survey Committee on the Problem of Ageing and the Care of Older People*. London: Nuffield Foundation.

Rowntree, S. and Lavers, G. (1951) *English Life and Leisure*. London: Macmillan and Company.

Royal Commission on Long Term Care (1999) *With Respect to Old Age: Long Term Care – Rights and Responsibilities*. London: The Stationery Office.

Royal Commission on Population (1949) Report of the Select Committee on Public Service and Demographic Change. London: HMSO.

Royal Commission on the Aged Poor (1898) *Report of the Royal Commission on the Aged Poor, appointed to consider whether any alterations in the system of Poor Law Relief are desirable, in the case of persons whose destitution is occasioned by incapacity for work resulting from old age, or whether assistance could otherwise be afforded in those cases*. London: HMSO.

Rudnicka, E., Napierała, P., Podfigurna, A., Męczekalski, B., Smolarczyk, R. and Grymowicz, M. (2020) The World Health Organization (WHO) approach to healthy ageing. *Maturitas*, 139:6–11.

Ruspini, E. and Del Greco, M. (2017) Multigenerational tourism. In L. Lowry (ed.), *The Sage International Encyclopedia of Travel and Tourism*. London: SAGE, 849–51.

Ryu, E., Hyun, S. and Shim, C. (2015) Creating new relationships through tourism: a qualitative analysis of tourist motivations of older individuals in Japan. *Journal of Travel & Tourism Marketing*, 32(4), 325–38.

Sánchez-González, D., Rojo-Pérez, F., Rodríguez-Rodríguez, V. and Fernández-Mayoralas, G. (2020) Environmental and psychosocial interventions in age-friendly communities and active ageing: a systematic review. *International Journal of Environmental Research and Public Health*, 17(22):1–35.

Sanderson, W. and Scherbov, S. (2015) Faster increases in human life expectancy could lead to slower population aging. *PLOS One*, 10(4):e0121922.

Sanford, C. (2002) Pre-travel advice: an overview. *Primary Care – Clinics in Office Practice*, 29(4):767–85.

Sangkharat, K., Mahmood, M., Thornes, J., Fisher, P. and Pope, F. (2020) Impact of extreme temperatures on ambulance dispatches in London, UK. *Environmental Research*, 182.

Sangpikul, A. (2008a) A factor-cluster analysis of tourist motivations: A case of US senior travelers. *Tourism*, 56(1), 23–40.

Sangpikul, A. (2008b) Travel motivations of Japanese senior travellers to Thailand. *International Journal of Tourism Research*, 10(1): 81–94.

Schäfer, D. (2011) *Old Age and Diseases in Early Modern Medicine*. London: Routledge.

Scharf, T. and Keating, N. (2012) Social exclusion in later life a global challenge. In T. Scharf and N. Keating (eds.), *From Exclusion to Inclusion in Old Age*. Bristol: Bristol University Press, 1–16.

Scheid, T. and Brown, T. (2010) *A Handbook for the Study of Mental Health* (2nd edition). New York: Cambridge University Press.

Scherger, S., Nazroo, J. and Higgs, P. (2010) Leisure activities and retirement: Do structures of inequality change in old age? *Ageing and Society*, 31(1):146–72.

Schewe, C. (1988) Marketing to our ageing population. *Journal of Consumer Marketing*, 5(3): 61–73.

Schiffman, L. and Sherman, E. (1991) Value orientations of new age elderly: The coming of an ageless market. *Journal of Business Research*, 22 (2): 187–94.

Schor, J. (1993) *Overworked American: The Unexpected Decline of Leisure*. New York: Basic Books.

Schulz, J. H. (2001) *The Economics of Aging*. Westport, CT: Greenwood Publishing Group.

Schumpeter, J. (1909) On the concept of social value. *The Quarterly Journal of Economics*, 23(2): 213–32.

Schwiter, K., Berndt, C. and Truong, J. (2018) Neoliberal austerity and the marketisation of elderly care. Social and Cultural Geography, 19(3), 379–99.

Scott, A. and Gratton, L. (2020) *The New Long Life: A Framework for Flourishing in a Changing World*. London: Bloomsbury Publishing.

Scott, J. (2012) Tourism, civil society and peace in Cyprus. *Annals of Tourism Research*, 39(4):2114–32.

Sedgley, D., Pritchard, A. and Morgan, N. (2006) Understanding older women's leisure: the value of biographical research methods. *Tourism*, 54(1):43–51.

Sedgley, D., Pritchard, A. and Morgan, N. (2011) Tourism and ageing: A transformative research agenda. *Annals of Tourism Research*, 38(2):422–36.

Sedgley, D., Pritchard, A., Morgan, N. and Hanna, P. (2017) Tourism and autism: journeys of mixed emotions. *Annals of Tourism Research*, 66:14–25.

Seeman M. (2016) Travel risks for those with serious mental illness. *International Journal of Travel Medicine and Global Health*, 4(3):76–8.

Segovia-San-Juan, A., Saavedra, I. and Fernández-de-Tejada, V. (2017) Analyzing disability in socially responsible companies. *Social Indicators Research*, 130(2): 617–45.

Sellick, M. (2004) Discovery, connection, nostalgia: key travel motives within the senior market. *Journal of Travel & Tourism Marketing*, *17*(1): 55–71.

Sessa, A. (1983) *Elements of Tourism*. Rome: Cantal.

Sevilla, A., Gimenez-Nadal, J. and Gershuny, J. (2012) Leisure inequality in the United States: 1965–2003. *Demography*, 49(3):939–64.

Shanas, E. (1971) The sociology of aging and the aged. *The Sociological Quarterly*, 12(2): 159–76.

Sharpless, N. (2018) *The Challenging Landscape of Cancer and Aging: Charting a Way Forward.* www.cancer.gov/news-events/cancer-currents-blog/2018/sharpless-aging-cancer-research

Shaw, G. and Coles, T. (2004) Disability, holiday making and the tourism industry in the UK: a preliminary survey. *Tourism Management*, 25(3):397–403.

Sheller, M. and Urry, J. (2006) The new mobilities paradigm. *Environment and Planning A: Economy and Space*, 38(2):207–26.

Sherman, S. (1974) Leisure activities in retirement housing. *Journals of Gerontology*, 29(3):325–35.

Shi, Y. (2006) The accessibility of Queensland visitor information centres' websites. *Tourism Management*, 27(5):829–41.

Shock, N. (1952) *Trends in Gerontology*. Stanford, CA: Stanford University Press.

Shock, N. (2020) Human aging: physiology and sociology. *Britannica*. www.britannica.com/science/human-aging.

Shoemaker, S. (1989) Segmentation of the senior pleasure travel market. *Journal of Travel Research*, 27(3):14–21.

Shoemaker, S. (2000) Segmenting the mature market: 10 years later. *Journal of Travel Research*, 39(1):11–26.

Shostack, L. (1984) Designing services that deliver. *Harvard Business Review*, 62(1):133–9.

Sie, L., Patterson, I. and Pegg, S. (2016) Towards an understanding of older adult educational tourism through the development of a three-phase integrated framework. *Current Issues in Tourism*, 19 (2):100–36.

Siikamäki, H., Kivelä, P., Fotopoulos, M., Ollgren, J. and Kantele, A. (2015) Illness and injury of travellers abroad: Finnish nationwide data from 2010 to 2012, with incidences in various regions of the world. *Eurosurveillance*, 20(19):21128.

Silcock, D. (2015) *Challenge for the Retirement Income Market over the next few Decades*. London: Government Office for Science.

Silverstein, N., Garcia, C. and Landis, A. (2001) Museums and aging: reflections on the aging visitor, volunteer, and employee. *Journal of Museum Education*, 26(1):3–7.

Singleton, J. (2017) Foreword: role of leisure and physical activity for Millennials, Generation X (Seniors/Older Adult in Training), and current cohort of seniors. *Topics in Geriatric Rehabilitation*, 33(3):153–5.

Siren, A. and Haustein, S. (2015) How do baby boomers' mobility patterns change with retirement? *Ageing and Society*, 36(5):988–1007.

Sjöberg, L. (2018) Using a life-course approach to better understand depression in older age. (Unpublished PhD, The Aging Research Center (ARC) Department of Neurobiology, Care Sciences and Society.)

Skinner, M. W., Andrews, G. J. and Cutchin, M. P. (2017) *Geographical Gerontology: Perspectives, Concepts, Approaches*. London: Routledge.

Skinner, M. W., Cloutier, D. and Andrews, G. J. (2014) Geographies of ageing: progress and possibilities after two decades of change. *Progress in Human Geography*, 39(6):776–99.

Small, J. (2003) The voices of older women tourists. *Tourism Recreation Research*, 28(2):31–9.

Small, J., Darcy, S. and Packer, T. (2012) The embodied tourist experiences of people with vision impairment: management implications beyond the visual gaze. *Tourism Management*, 33(4):941–50.

Small, J. and Harris, C. (2012) Obesity and tourism: rights and responsibilities. *Annals of Tourism Research*, 39(2):686–707.

Smith, A. and Graham, A. (eds.) (2019) *Destination London: The Expansion of the Visitor Economy*. London: University of Westminster Press.

Smith, S. and House, M. (2006) Snowbirds, sunbirds, and stayers: seasonal migration of elderly adults in Florida. *The Journals of Gerontology: Series B*, 61(5):S232 – S239.

Soares-Miranda, L., Siscovick, D., Psaty, B., Longstreth, W. and Mozaffarian, D. (2016) Physical activity and risk of coronary heart disease and stroke in older adults. *Circulation*, 133(2):147–55.

Soja, E. (2010) *Seeking Spatial Justice*. Minnesota: University of Minnesota Press.

Sprod, J., Ferrar, K., Olds, T. and Maher, C. (2015) Changes in sedentary behaviours across the retirement transition: a systematic review. *Age and Ageing*, 44(6):918–25.

Stacey, T-L., Froude, E., Trollor, J. and Foley, K-R. (2019) Leisure participation and satisfaction in autistic adults and neurotypical adults. *Autism*, 23(4):993–1004.

Stebbins, R. (1982) Serious leisure: a conceptual statement. *Pacific Sociological Review*, 25(2):251–72.

Stebbins, R. (1992) *Amateurs, Professionals, and Serious Leisure*. Montreal: McGill-Queens University Press.

Stebbins, R. (1998) *After Work: The Search for an Optimal Leisure*. Calgary: Detselig Enterprises Lifestyle

Stebbins, R. (2015) *Between Work and Leisure: The Common Ground of Two Separate Worlds*. New Brunswick, NJ: Transaction Publishers.

Steels, S. (2015) Key characteristics of age-friendly cities and communities: a review. *Cities*, 47:45–52.

Steeman, E., De Casterlé, B., Godderis, J. and Grypdonck, M. (2006) Living with early-stage dementia: a review of qualitative studies. *Journal of Advanced Nursing*, 54(6):722–38.

Stephens, C., and Breheny, M. (2018) *Healthy Ageing: A Capability Approach to Inclusive Policy and Practice*. London: Routledge.

Stiglitz, J. (2020) *People, Power and Profits: Progressive Capitalism in an Age of Discontent*. London: Penguin.

Stockdale, J. (1987) *Methodological Techniques in Leisure Research*. London: Sports Council and ESRC.

Strain, L., Grabusic, C., Searle, M. and Dunn, N. (2002) Continuing and ceasing leisure activities in later life: a longitudinal study. *Gerontologist*, 42:217–23.

Strauss-Blasche, G., Ekmekcioglu, C. and Marktl, W. (2000) Does vacation enable recuperation? Changes in well-being associated with time away from work. *Occupational Medicine*, 50(3):167–72.

Strauss-Blasche, G., Reithofer, B., Schobersberger, W, Ekmekcioglu, C. and Marktl, W. (2005) Effect of vacation on health: moderating factors of vacation outcome. *Journal of Travel Medicine*, 12(2):94–101.

Streib, G. and Thompson, W. (eds.) (1958) Adjustment in retirement. *The Journal of Social Issues*, 14(2):1–64.

Streltzer, J. (1979) Psychiatric emergencies in travelers to Hawaii. Comprehensive Psychiatry, 20(5):463–8.

Stroud, D. (2005) *The 50 Plus Market*. Philadelphia: Kogan Page.

Stroud, D. and Walker, K. (2013) *Marketing to the Ageing Consumer*. Basingstoke: Palgrave Macmillan.

Stuart-Hamilton, I. (2006) *The Psychology of Ageing: An Introduction*. London: Jessica Kingsley.

Sun Life (2019) *Ageist Britain*. www.sunlife.co.uk/siteassets/documents/ageist-report-2019.pdf

Sun Life (2020) *Retiring Ageism*. www.sunlife.co.uk/over-50-life-insurance/over-50-data-centre/ageism/

Sunstein, C. (2014) *Why Nudge? The Politics of Liberal Paternalism*. New Haven, CT: Yale University Press.

Surdam, D. (2015) *Century of the Leisured Masses*. Oxford: Oxford University Press.

Swift, H. and Steeden, B. (2020) *Doddery But Dear: Examining Age-Related Stereotypes*. London: Centre for Ageing Better.

Tabloski, P. A. (2004) Global aging: Implications for women and women's health. *Journal of Obstetric, Gynecologic, and Neonatal Nursing*, 33(5):627–38.

Tapper, R. and Font, X. (2004) *Tourism Supply Chains, Report of a Desk Research Project for the Travel Foundation*. Leeds: Leeds Metropolitan University.

Taylor, F. (1911) *The Principles of Scientific Management*. New York: Harper & Brothers.

Tecau, A., Bratucu, G., Tescaşiu, B., Chiţu, I., Constantin, C. and Foris, D. (2019) Responsible tourism-integrating families with disabled children in tourist destinations. *Sustainability*, 11(16): www.mdpi.com/2071-1050/11/16/4420

Teddlie, C., Tashakkori, A., Johnson, R. (2020) *Foundations of Mixed Methods Research: Integrating Quantitative and Qualitative Approaches in the Social and Behavioral Sciences*. Thousand Oaks, CA: SAGE.

Thaler, R. and Sunstein, C. (2008) *Nudge: Improving Decisions about Health, Wealth and Happiness*. New Haven, CT: Yale University Press.

Thane, P. (1989) History and the sociology of ageing. *Social History of Medicine*, 2(1):93–6.

The Elder (2020) Live-in care agency finds 1 in 3 elderly people more lonely in wake of COVID-19. www.elder.org/the-elder/survey-on-elderly-loneliness/.

The National Autistic Society and Visit England (2018) *Welcoming Autistic People: A Guide for Tourism Venues*. London: Visit England.

Thomas, C. and Milligan, C. (2015) *How Can and Should UK Society Adjust to Dementia?* York: Joseph Rowntree Foundation Viewpoint Paper.

Thompson, A. (1949) Problems of ageing and chronic sickness. *British Medical Journal*, 30: 243–50.

Thompson, P. (1992) 'I don't feel old': subjective ageing and the search for meaning in later life. *Ageing and Society*, 12(1):23–47.

Thomson, D, (1984) The decline of social welfare: falling state support for the elderly since early Victorian times. *Ageing and Society*, 4(4):429–49.

Thurm, A. and Swedo, S. (2012) The importance of autism research. *Dialogues in Clinical Neuroscience*, 14(3):219–22.

Tibbits, C. (ed.) (1960) *Handbook of Social Gerontology: Societal Aspects of Ageing*. Chicago: The University of Chicago Press.

Timonen, V. (2016) *Beyond Successful and Active Ageing: A Theory of Model Ageing*. Bristol: Policy Press.

Tinetti, M., Huang, A. and Molnar, F. (2017) The Geriatrics 5M's: a new way of communicating what we do. *Journal of the American Geriatric Society*, 65(9):2115.

Tint, A., Thomson, K. and Weiss, J. (2017) A systematic literature review of the physical and psychosocial correlates of Special Olympics participation among individuals with intellectual disability. *Journal of Intellectual Disability Research*, 61(4):301–24.

Toepoel, V. (2013) Ageing, leisure, and social connectedness: how could leisure help reduce social Isolation of older people? *Social Indicators Research*, 113(1):355–72.

Togunu-Bickersteth, F. (1987) Chronological definitions and expectations of old age among young adults in Nigeria. *Journal of Aging Studies*, 1(2):113–24.

Togunu-Bickersteth, F. (1988) Perception of old age among Yoruba aged. *Journal of Comparative Family Studies*, 19(1):113–23.

Tomljenovic, R. and Faulkner, B. (2000) Tourism and older residents in a Sunbelt Resort. Annals of Tourism Research, 27(1), 93–114.

Tornstam, L. (2011) Maturing into gerotranscendence. *The Journal of Transpersonal Psychology*, 43(2):166–80.

Torrington, J. (2014) *What Developments in the Built Environment will Support the Adaptation and 'Future Proofing' of Homes and Local Neighbourhoods so that People Can Age Well in Place over the Life Course, Stay Safe and Maintain Independent Lives? Future of an Ageing Population: Evidence Review.* N.p.: Foresight/Government Office for Science. https://assets.publishing.service.gov.uk/government/uploads/system/uploads/attachment_data/file/445583/gs-15-11-future-ageing-homes-neighbourhoods-er21.pdf.

Tourism Australia (2010) No leave no life. www.tourism.australia.com/en/about/our-campaigns/past-campaigns/no-leave-no-life.html.

Tourism Toronto Partners (2019) *Toronto's Visitor Economy.* www.destinationtoronto.com/research/business-intelligence/visitor-economy-study/

Townsend, P. (1957) *The Family Life of Old People: An Inquiry in East London.* Glencoe, IL: The Free Press.

Townsend, P. (1979) *Poverty in the United Kingdom.* Harmondsworth: Penguin.

Townsend, P. (1981) The structured dependency of the elderly: a creation of social policy in the twentieth century. *Ageing and Society* 1 (1):5–28.

Trapper, J. (2019) All under one roof: The rise and rise of multigenerational life. *The Guardian*, 10 May.

Tretheway, M. and Mak, D. (2006) Emerging tourism markets: ageing and developing economies. *Journal of Air Transport Management*, 12(1):21–7.

Tsartsara, S. (2018) Definition of a new type of tourism niche – The geriatric tourism. *International Journal of Tourism Research*, 20(6):796–9.

Turner, N. and Cannon, S. (2018) Aligning age-friendly and dementia-friendly communities in the UK. *Working with Older People*, 22(1):9–19.

Turner, R., Miller, G. and Gilbert, D. (2001) The role of UK charities and the tourism industry. *Tourism Management*, 22(5), 463–72.

Tutuncu, O. and Lieberman, L. (2016) Accessibility of hotels for people with visual impairments: from research to practice. *Journal of Visual Impairment and Blindness*, 110(3):163–75.

UK Cabinet Office Performance and Innovation Unit (2001) *A Futurist's Toolbox: Methodologies in Futures Work, Strategic Futures Team.* London: UK Cabinet Office Performance and Innovation Unit.

United Nations Department of Economic and Social Affairs (2016) *Income Poverty in Old Age: An Emerging Development Priority.* New York: United Nations.

United Nations Department of Economic and Social Affairs (2020) *World Population Ageing 2019*, www.un.org/en/development/desa/population/publications/pdf/ageing/WorldPopulationAgeing2019-Report.pdf.

United Nations Division for Social Policy and Development (1998) *The Ageing of the World's Population.* New York: United Nations. www.un.org/esa/socdev/agewpop.htm.

United Nations Educational, Scientific, and Cultural Organization. (n.d.). MOST Clearing House: Best practices. www.unesco.org/most/bphome.htm#1

United States Federal Security Agency (1951) *Man and His Years*. Raleigh, NC: Publications Institute.

UNWTO (2010) *Demographic Change and Tourism*. Madrid: UNWTO.

UPIAS (Union of the Physically Impaired Against Segregation) (1976) *Fundamental Principles of Disability*. London: UPIAS.

Urry, J. (2007) *Mobilities*. London: Polity.

Valk T. (2017) Psychiatric issues in travel medicine: what is needed now. *Journal of Travel Medicine*, 24(5):tax037.

Var, T., Yeşiltaş, M., Yayli, A. and Öztürk, Y. (2011) A study on the travel patterns of physically disabled people. *Asia Pacific Journal of Tourism Researc*h, 16(6), 599–618.

Veblen, T. (1953) *The Theory of the Leisure Class*. New York: The New York American Library (originally published 1899).

Veenhoven, R. and Hagerty, M. (2006) Rising happiness in nations 1946–2004: a reply to Easterlin. *Social Indicators Research*, 79:421–36.

Verbrugge, L. (2016) Disability experience and measurement. *Journal of Aging and Health*, *28*(7), 1124–58.

Verbrugge, L. and Jette, A. (1993) The disablement process. *Social Science and Medicine*, 6(1):1–14.

Viallon, P. (2012) Retired snowbirds. *Annals of Tourism Research*, 39(4): 2073–91.

Viry, G. and Kaufmann, V. (eds.) (2015) *High Mobility in Europe: Work and Personal Life*. Basingstoke: Palgrave Macmillan.

VisitEngland (2019) *One Minute to Midnight*. London: VisitEngland.

Von Soest, T., Luhmann, M., Hansen, T. and Gerstorf, D. (2020) Development of loneliness in midlife and old age: its nature and correlates. *Journal of Personality and Social Psychology*, 118:388–406.

Vos, L. and Page, S. J. (2020) Marketization, performative environments, and the impact of organizational climate on teaching practice in business schools. *Academy of Management Learning and Education*, 19(1):59–80.

Vos, T., Barber, R. M., Bell, B., Bertozzi-Villa, A., Biruyukov, S., Bollinger, I., … Murray, C. J. (2013) Global, regional, and national incidence, prevalence, and years lived with disability for 301 acute and chronic diseases and injuries in 188 countries, 1990–2013: a systematic analysis for the Global Burden of Disease study. *The Lancet*, 386(9995):743–800.

Wæhrens, E., Brandt, Å., Peoples, H. and la Cour, K. (2020) Everyday activities when living at home with advanced cancer: a cross-sectional study. *European Journal of Cancer Care*, 29(5):e13258.

Wahl, H. (2006) Introduction: The person – environment perspective in ageing research. In H. Wahl, H. Brenner, D. Rothenbacher and C. Rott (eds.), *The Many Faces of Health, Competence and Well-Being in Old Age*. Dordrecht: Springer, 3–6.

Walker, A. (1980) The social creation of poverty and dependency in old age. *Journal of Social Policy*, 9 (1):49–75.

Walker, A. (1981) Towards a political economy of old age. *Ageing and Society*, 1(1):73–94.

Walker, A. (2015) The concept of active ageing. In A. Walker and C. Aspalter (eds.), *Active Ageing in Asia*. London: Routledge, 14–29.

Walker, A. (ed.) (2018) *The Future of Ageing in Europe*. Singapore: Palgrave Macmillan.

Walker, A. and Aspalter, C. (eds.) (2015) *Active Ageing in Asia*. London: Routledge.

Walton, J. (1983) *The English Seaside Resort: A Social History, 1750–1914*. Leicester: Leicester University Press.

Walton, J. (2000) *The British Seaside: Holidays and Resorts in the Twentieth Century*. Manchester: Manchester University Press.

Walton, J. (ed.) (2014) *Mineral Spring Resorts in Global Perspective*. London: Routledge.

Walton, K. M. (2019) Leisure time and family functioning in families living with autism spectrum disorder. *Autism*, 23(6):1384–97.

Wan, W. and Antonucci, T. (2016) Social exchange and aging. In N. Pachana (ed.), *Encyclopedia of Geropsychology*. Singapore: Springer, 1–9.

Wang, K. and Feng, S. (2014) Exploring the travel service demands of elderly people. *Gerontechnology*, 13(2):299.

Wang, W., Wu, W., Luo, J. and Lu, J. (2017) Information technology usage, motivation, and intention: A case of Chinese urban senior outbound travelers in the Yangtze River Delta region. *Asia Pacific Journal of Tourism Research*, 22(1): 99–115.

Ward, A. (2014) Segmenting the senior tourism market in Ireland based on travel motivations. *Journal of Vacation Marketing*, 20(3): 267–77.

Warnes, A. (1981) Towards a geographical contribution to gerontology. *Progress in Geography*, 5(3), 317–41.

Warnes, A. (1982) Geographical perspectives on ageing. In A. M. Warnes (ed.), *Geographical Perspectives on the Elderly*, New York: John Wiley, 1–31.

Warnes, A. (1990) Geographical questions in gerontology: needed directions for research. *Progress in Human Geography*, 14(1):24–56.

Warnes, A. (2001) The international dispersal of pensioners from affluent countries, *International Journal of Population Geography*, 7(6):373–88.

Warnes, A. (2009) Ageing and Mobility. In R. Kitchin and N. Thrift (Eds.), *International Encyclopedia of Human Geography*. Oxford: Elsevier, 36–41.

Warnes, A. and Law, C. (1984) The elderly population of Great Britain: Locational trends and policy implications. *Transactions of the Institute of British Geographers* 9:37–59.

Waterman, R., Peters, T. and Phillips, J. (1980) Structure is not organization. *Business Horizons*, 23(3):14–26.

Watson, N., Roulstone, A. and Thomas, C. (eds.) (2016) *Routledge Handbook of Disability Studies*. London: Routledge.

Weiner, B. (2009) A theory of organizational readiness for change. *Implementation Science*, 4(1):67.

Weiss, R. S. (1975) *Loneliness: The Experience of Emotional and Social Isolation*. Massachusetts: MIT Press.

Wen, J., Huang, S. and Goh, E. (2020) Effects of perceived constraints and negotiation on learned helplessness: a study of Chinese senior outbound tourists. *Tourism Management*, 78, 104059.

Wenger, E. (1998) *Communities of Practice: Learning, Meaning and Identity*. Cambridge: Cambridge University Press.

Wenger, E. and Snyder, W. (2000) Communities of practice: the organizational frontier. *Harvard Business Review*, 78(1):139–45.

Weston, R. (1996) Have fun in the sun: Protect yourself against skin damage. In S. Clift and S. J. Page (eds.), *Health and the International Tourist*, 235–59. London: Routledge.

WHO (World Health Organization) (1992) *International Statistical Classification of Impairments, Activities and Participation (ICIDH-2)*. Geneva: World Health Organization.

WHO (World Health Organization) (1994) *Health for All*. Geneva: World Health Organization.

WHO (World Health Organization) (2001) *International Classification of Functioning, Disability and Health (ICF)*. Geneva: World Health Organization.

WHO (World Health Organization) (2002) Proposed working definition of an older person in Africa for the MDS Project. www.who.int/healthinfo/survey/ageingdefnolder/en/#:~:text=Proposed%20working%20definition%20of%20an%20older%20person%20in%20Africa%20for%20the%20MDS%20Project,-Note%3A%3A%20This%20paper&text=As%20far%20back%20as%201875,or%2065%20years%20for%20eligibility

WHO (World Health Organization) (2007a) *Global Report on Falls Prevention in Older Age*. Geneva: World Health Organization.

WHO (World Health Organization) (2007b) *Global Age Friendly Cities*: *A Guide*. Geneva: World Health Organization.

WHO (World Health Organization) (2009) *Public Health Campaigns: Getting the Message Acr*oss. Geneva: World Health Organization.

WHO (World Health Organization) (2010) *Global Recommendations on Physical Activity for Health: Physical Activity and Older Adults*. Geneva: World Health Organization. www.who.int/publications/i/item/9789241599979

WHO (World Health Organization) (2012) *Dementia: A Public Health Pri*ority. Geneva: Alzheimer's Disease International and World Health Organisation.

WHO (World Health Organization) (2018) *Ageing and Health*. Geneva: World Health Organization. www.who.int/news-room/fact-sheets/detail/ageing-and-health.

WHO (World Health Organization) (2019) *World Report of Vision*. Geneva: World Health Organization.

WHO (World Health Organization) (2020) *Depression*. Geneva: World Health Organization. www.who.int/news-room/fact-sheets/detail/depression.

WHO (World Health Organization) Assessment, Classification and Epidemiology Unit. (1999)ʹ *International Classification of Functioning and Disability: ICIDH-2*, Geneva: World Health Organization.

Whyte, C. and Fortune, D. (2017) Natural leisure spaces in long-term care homes: challenging assumptions about successful aging through meaningful living. *Annals of Leisure Research*, 20(1):7–22.

Wild, S., Roglic, G., Green, A., Sicree, R. and King, H. (2004) Global prevalence of diabetes: estimates for the year 2000 and projections for 2030. *Diabetes Care*, 27: 1047–53.

Wilkinson, C. and Ripley, D. (2020) Travelling with heart disease: restricted choice and restrictive cost. *Journal of the Royal College of Physicians of Edinburgh*, 50(3):222–3.

Wilks, J. and Page, S. J. (eds.) (2003) *Managing Tourist Health and Safety*. Oxford: Pergamon.

Williams, A. (1998) Therapeutic landscapes in holistic medicine. *Social Science & Medicine*, 46(9):1193–1203.

Williams, A. M. and Hall, C. M. (2000) Tourism and migration: new relationships between production and consumption. *Tourism Geographies*, 2(1):5–27.

Wilson, N. (2017) Sea sickness: cruising towards a healthy holiday. *Australian Journal of Pharmacy*, 98(1161):81–3.

Wolfe, R. J. (1966) Recreational travel: the new migration. *Geographical Bulletin*, 9:73–9.

Wood, E., Jepson, A. and Stadler, R. (2018) Understanding the well-being potential of participatory arts events for the over 70s: a conceptual framework and research agenda. *Event Managem*ent, 22(6):1083–1101.

World Bank (n.d.) Civil society: Overview. www.worldbank.org/en/about/partners/civil-society/overview

World Economic Forum (2015) *How 21st Century Longevity Can Create Markets and Drive Economic Growth*. Geneva: World Economic Forum.

World Values Survey (2016) *World Value Survey*. www.worldvaluessurvey.org/wvs.jsp.

Wright, S. (ed.) (2016) *Autism Spectrum Disorder in Mid and Later Life*. Philadelphia, PA: Jessica Kingsley Publishers,

Wu, H., Mach, J., Le Couteur, D. and Hilmer, S. (2020) Fall-related mortality trends in Australia and the United Kingdom: implications for research and practice. *Maturitas*, 142:68–72.

Yachin, M. and Nimrod, G. (2021) Innovation in later life: a study of grandmothers and Facebook. *The International Journal of Aging and Human Development*, 92(4).

Yamagishi, T., Kamiya, H., Kakimoto, K., Suzuki, M. and Wakita, T. (2020) Descriptive study of COVID-19 outbreak among passengers and crew on Diamond Princess cruise ship, Yokohama Port, Japan, 20 January to 9 February 2020. *Euro Surveillance*, 25(23):pii=2000272.

Yang, K. and Victor, C. (2011) Age and loneliness in 25 European nations. *Ageing and Society*, 31(8):1368–88.

Yau, M., McKercher, B. and Packer, T. (2004) Traveling with a disability – More than an Access Issue. *Annals of Tourism Research*, 31(4):946–60.

Yeh, C., Cheng, H. and Shi, S. (2018) Public–private pension mixes in East Asia: institutional diversity and policy implications for old-age security. *Ageing and Society*, 40(3):604–25.

Yeoman, I., Hsu, C., Smith, K and Watson, S. (2010) *Tourism and Demography*. Oxford: Goodfellow Publishers.

Yiannakis, A. and Gibson, H. (1992) Roles tourists play. *Annals of Tourism Research*, 19(2):287–303.

Yilmaz, N. and Karaca, S. (2020) Dissatisfaction with life and absence of leisure time activity. *Psychogeriatrics*, 20(3):337–44.

Yim, J. (2016) Therapeutic benefits of laughter in mental health: a theoretical review. *Tohoku Journal of Experimental Medicine*, 239(3):243–9.

Ylänne, V. (2016) Too old to parent? Discursive representations of late parenting in the British press. *Discourse & Communication*, 10(2):176–97.

You, X. and O'Leary, J. (1999) Destination behaviour of older UK travellers. *Tourism Recreation Research*, 24(1):23–34.

Young, M. and Willmott, P. (1973) *The Symmetrical Family*. London: Routledge and Kegan Paul.

Yule, G. (1899) An investigation into the causes of changes in pauperism in England, chiefly during the last two intercensal decades (Part I). *Journal of the Royal Statistical Society*, 62(2):249–95.

Zaidi, A., Harper, S., Howse, K., Lamura, G. and Perek-Bialas, J. (eds.) (2018) *Building Evidence for Active Ageing Policies*. Singapore: Palgrave Macmillan.

Zajadacz, A. (2014) Sources of tourist information used by Deaf people. Case study: The Polish Deaf community. *Current Issues in Tourism*, 17(5):434–54.

Zamboni, M., Mazzali, G., Zoico, E., Harris, T. B., Meigs, J. B., Di Francesco, V., ... Bosello, O. (2005) Health consequences of obesity in the elderly: a review of four unresolved problems, *International Journal of Obesity*, 29:1011–29.

Zeithaml, V., Bitner, M. and Gremler, D. (2006) *Services Marketing: Integrating Customer Focus Across the Firm* (4th ed.). Boston, MA: McGraw-Hill Irwin.

Zeithaml, V. and Bitner, M. (1996) *Services Marketing*. New York: McGraw-Hill.

Zhao, Q., Li, Z. and Chen, T. (2016) The impact of public pension on household consumption: evidence from China's survey data. *Sustainability* 8(9):890.

Index

Numbers in *italics* refer to figures. Numbers in **bold** refer to tables.

Taylor & Francis Group
an **informa** business

Taylor & Francis eBooks

www.taylorfrancis.com

A single destination for eBooks from Taylor & Francis
with increased functionality and an improved user
experience to meet the needs of our customers.

90,000+ eBooks of award-winning academic content in
Humanities, Social Science, Science, Technology, Engineering,
and Medical written by a global network of editors and authors.

TAYLOR & FRANCIS EBOOKS OFFERS:

A streamlined
experience for
our library
customers

A single point
of discovery
for all of our
eBook content

Improved
search and
discovery of
content at both
book and
chapter level

REQUEST A FREE TRIAL
support@taylorfrancis.com

Routledge
Taylor & Francis Group

CRC Press
Taylor & Francis Group

For Product Safety Concerns and Information please contact our EU
representative GPSR@taylorandfrancis.com
Taylor & Francis Verlag GmbH, Kaufingerstraße 24, 80331 München, Germany

www.ingramcontent.com/pod-product-compliance
Lightning Source LLC
Chambersburg PA
CBHW060240220326
41598CB00027B/3995

* 9 7 8 1 0 3 2 0 7 2 9 0 6 *